Biodiversity and Sustainable Development

BIODIVERSITY AND SUSTAINABLE DEVELOPMENT

Dr. Reena Mohanka
Dr. M.P. Singh

Prints Publications Pvt Ltd
New Delhi

Published by

Prints Publications Pvt Ltd
Viraj Tower-2, 4259/3, Ansari Road,
Darya Ganj, New Delhi-110002
Tel. : +91-11-45355555
Fax: +91-11-23275542
E-mail : contact@printspublications.com
Website : www.printspublications.com

First Edition : 2022 (Hardbound)

ISBN: 978-93-936747-0-8

Price: ₹ 2495/-

Published and Printed by Mr. Pranav Gupta (Managing Director) on behalf of Prints Publications Pvt Ltd, New Delhi.

PREFACE

Mother Nature is an Integral part of the Indian culture and heritage, In Hindu mythological stories, rivers and Mountains emerge as angels, and air and trees appear as human beings, being able to feel, hear, and talk. Despite this "humanization" of the environment, the abuse and destruction of natural resources has occurred at an alarming rate.

In the Indian Constitution, several provisions have been given to protect environment and safeguarding the natural resources by adopting a New Forest Policy (1988), Water Policy (1987), Housing Policy (1988) and Land Use Policy (1987). It has also adopted "The National Conservatory, Strategy and PolicyStatement on Environment and Development (1992), Policy statement on the Abatement of Pollution (1992), Environmental Action Programme (1993) and Plethora of legislations".

India's stand on global environmental issue has been clearly articulated at several levels, both before and after the United Nations Conference on Environment and Development (UNCED) in the, 1992 at the Rio De Janeiro. Human beings are promised for a healthy and productive life in harmony with nature and the eradication of poverty, which is considered as an indispensable requirement for sustainable development.

It the field forest policy, the Indian Government has resorted to symbolism. Despite of the Forest Act (in existence since 1927), arid an extensive forest bureaucracy, the loss of forest area has been increasing rate.

In the workshop on a New Forest Policy, held at the Indian Social Institute in April, 1982, most participants identified the high cost of bureaucratic involvement, and realized the need for a more organized public support in order to design an effective forest management program.

The extinction of wildlife and plant and destruction of ecosystems have been the major concern of policymakers and environmentalists as without them civilization could be quite unimaginary. For the sake of human welfare, a new subject biotechnology has come into existence which involves the use of all life norms for the welfare of human beings as well as for protecting biodiversity using the scientific and engineering principles.

We should must develop property blend technological advancement with ancient wisdom of our sages to live in harmony with nature. The conservation and development of forest require the eradication of poverty and symbiotic relationship, with development at projects. Forests dwellers have to be convinced that they have stake in ecological conservation. It must be recognized that economic security tied with ecological security.

The main aim of this book is to integrate our concerns for conservation and development of Eco-environmental protection, poverty alleviation and human welfare. The book presents eco-friendly models for conservation of natural resources, protection, rehabilitation, afforestation and greening activities with vies to maintain and restore the ecological balance.

We hope this book will help in creating Eco environmental awareness and in the implementation of numerous Eco regeneration projects.

A number of experts have contributed to the development of this book. Their contribution is great fully acknowledged.

Reena Mohanka

M.P Singh

CONTENTS

1

BIODIVERSITY CONSERVATION: VALUE ADDITION THROUGH BIOTECHNOLOGY

Biotechnology is the third wave in biological sciences and represents such an interface of basic and applied sciences, where gradual and subtle transformation of science into technology can be witnessed. Biotechnology is also defined as the applications of scientific and engineering principles to the processing of material by biological agents to provide goods and services. United States Congress's Office of Technology Assessment defined biotechnology as any technique that used living organisms to make or modify a product, to improve plants or animals or to develop microorganisms for specific uses. The document focuses on the development and application of modern biotechnology based on new enabling techniques of recombinant-DNA technology, often referred to as genetic engineering.

Since biotechnology involves the use of all life forms for human welfare. Therefore, extinction of wild species and destruction of ecosystems has been a major concern of policy markers and environmentalists. A discussion on biotechnology involving biodiversity is relevant, because biodiversity is being utilized to provide genes from wild species for biotechnological exercises.

Several biotechnological tools are now available to tackle various specific problems and to enhance the potentials of grass covers such as feeding value, propagation and persistence. Basically, in range grasses, several tools of biotechnology, comprising endosperm and anther culture, somaclonal variation, protoplast culture and fusion, transformation, etc, can address breaking the apomictic barrier for recombination of desired traits. The grass biotechnologists are actively engaged to achieve this goal. Besides, the mapping of genes, isolation, cloning and characterization of important genes/ traits in range species as well as in other crops can ultimately improve the value of grass cover through genetic transformation in the pasture species. It is envisaged that by following these approaches anti-quality factors can be suppressed and the digestibility of grasses can be increased. The vast array of gene mapping technologies now allows even tracking the genes for polyphonic traits and the metric traits like seed production; the biomass would also be hopefully improved. The current extensive research on nitrogen fixation in the world's crop plants would certainly go a long way in adding to the fertility and diversity of grass covers. Propagation of specific genotypes of endangered but useful species on a large scale in a short time frame is possible through the modern biotechnological approaches.

Germplasm Facilities

Recognizing the need for sophisticated facilities for research and development and providing services, the following additional germplasm facilities have been set up:

1. The National Facility for Microbial type Culture Collection at the Institute of Microbial Technology, Chandigarh, with over 1,600 culture in its stock.
2. The National Facility on Blue-green Algal Collection at the Indian Agriculture Research Institute, with over 500 strains and several pure-cultures as well

as soil-based cultures, which have been supplied to farmers for production of bio-fertilizers.

3. The National Facility for Marine Cyanobactena at the Bharatidasan University, Tiruchirapalli, which is coordinating extensive surveys on the southern coast.

4. The National Facility for Plant Tissues Cultures Repository at NBPGR, New Delhi, which has undertaken in vitro conservation of germplasm (cell), tissue, organ for medium and long-term, particularly for those species for which conventional methods are inadequate. It has 850 accessions of crop species and employs molecular methods of characterization and classification.

5. The National Facility for Laboratory Animals at the Central Drug Research Institute, Lucknow and the National Institute of Nutrition, Hyderabad have made available quality animals for biomedical research and industry in the country.

6. The National Facility for Animal Tissue and Cell Culture, Pune, an autonomous institution under Department of Biotechnology (DBT) has 1127 stock cultures comprising 594 different cell strains. The facility has supplied 401 culture consignments to 84 institutions throughout the country. It also has 50 rectors, plasmids and genomic libraries.

7. Three National Gene Banks for Medicinal and Aromatic Plants at the Central Institute of Medicinal and Aromatic Plants, Lucknow and NBPGR, New Delhi, for the northern region; and the Tropical Botanical Garden and Research Institute, Trivandrum, for peninsular India have been established. These banks will conserve important species of proven medicinal value, which are categorised as endangered, threatened or rare and are used extensively in traditional systems of medicine,

difficult to propagate, have significance for R&D for the future, and are of commercial values. India is the regional coordinator for Asia and also the overall coordinator for the establishment of Gene Banks of Medicinal and Aromatic Plants among G-15 countries.

Conservation Of Germplasm

Grass covers not only contain grasses but also a wealth of other potential herbaceous and woody species of socio-economic importance, including wild relative of important crop species. These valuable resources are today getting eroded at an alarming rate due to the loss of natural ecosystems. Therefore, rehabilitation and restoration of damaged ecosystem and the recovery of threatened and endangered species assumes greater importance. Traditional knowledge in terms of use of resources should be the key to maintenance of diversity in equilibrium. Participation of inhabitants in view of assured economic and commercial benefits would also ensure biodiversity conservation.

Both in situ and ex situ, the ICFRE and ICAR Institutions have put conservation methods into practice. In situ conservation is preferred option in case of grassland species because it allows for continuing adaptation of wild populations and the natural evolutionary processes. Following the man and biosphere approach advocated by the UNESCO, 14 sites have been identified as Biosphere Reserves. They have been set up in different parts of the country starting with Nilgiri biosphere reserve in 1986 covering an area of 5520 sq. km. Each biosphere reserve is intended to fulfil three complimentary functions, namely:

1. **Conservation function**, to preserve genetic diversity, ecosystem and landscapes.
2. **Development function**, to foster sustainable economic and human development.
3. **Logistic support function**, to support demonstration projects, environmental education and training,

research and monitoring related to local, national and global issues of conservation and sustainable development.

Genetic resources are renewable, provided they are well managed. Conservation and management of biodiversity, closely linked to the conservation and wise use of natural ecosystem, is of major importance for sustainable development. The tropical and sub-tropical areas are rich in term of plant genetic resources. India is particularly rich in biodiversity, including the ecosystem, species and genetic diversity, species and genetic diversity, by virtue of its varied climates and physical features.

In view of the unique importance of the grass covers in terms of total agro biodiversity scenario, it is appropriate to the realistic and assume the following as priority concerns in the future context. The updating of grass cover/grazing resources of India is required to be done on priority. Almost three decades have passed when grass cover of India was documented. The environmental degradation has caused enormous impact not only on the flora and fauna but the area represented activity. Remote sensing and GIS approach have been refined and made more widely available in the country during this period. These can be suitably used for the required updating.

1. Collection, evaluation and exploitation of promising grasses/legumes, which have not yet been tested for their prudential as forage species.
2. Inventory of grazing routes, grazing system and designing sustainable production system for migratory grazers of Himalayas as well as of the Thar Desert.
3. Soil and water conservation may be taken up an integral part of grassland development.
4. Greater thrust is required to be made for selection of native types from stress prone habitats i.e., ravines, riverine, alkaline, acidic, hot desert, cold

desert and steep slopes etc., their characterization, conservation and value addition.

5. In situ conservation of genetic wealth of above mentioned stress prone areas through botanical gardens can help in preserving several species which are facing extinction are less important today but may be of much value in future.

Ever since primitive man learned the art of farming and realized the economic utility of plants, he started saving selected seeds or vegetative propagules from one season to the next. Infact, he was practicing a type of germiplasm conservation and management. The primitive cultivars and their wild relatives constitute a poll of genetic diversity (germplasm or gene bank) which is invaluable for breeding programmes. However, the population and needs of man have steadily increased over time leading to persistent exploitation of natural resources. The pressure thus created led to the way for gradual exclusion of traditionally used genotypes. In this way, some of the valuable gene pools might be lost unless coordinated efforts are made towards the conservation of genetic stocks all over the world. Realizing the danger of erosion of genetic resources, the objective of providing necessary support to collection, conservation and utilisation of the plant genetic resources is concern of the world now.

Germplasm Conservation Using Cell And Tissue Culture

Objectives

1. To ensure the maximum level of genetic stability of the conserved material that is consistent with practical storage periods and with the available technology and resources.

2. To maintain accurate and detailed records concerning the biology of the conserved material, and its known history prior to collection. It is important to know the disease status of the material, and its full culture history including media details, the use of growth regulators, frequency of transfer etc. Cultures

and derived plants should be characterised at some level of morphology performance and any known genetic characteristics, including analysis of gene products and DNA fingerprinting where this is possible and appropriate. There is a responsibility to disseminate this information as widely as possible.

3. To develop and improve technology at all levels, and to increase the efficiency and utility of in vitro techniques in the conservation of plant germplasm.
4. To construct and maintain a policy for the in vitro conservation of those plant genetic resources appropriate to the collection with due regards to International, National and Regional/Local priorities.
5. Conservation of somaclonal and gemetoclonal variations in cultures.
6. Maintenance of recalcitrant seeds.
7. Conservation of cell lines producing medicines.
8. Storage of pollen for enhancing longevity.
9. Conservation of rare germplasm arising through somatic hybridization or other methods of genetic manipulations.
10. Delaying the process of ageing.
11. Storage of meristem culture for micro propagation, micro grafting and production of disease free plants.

Application of In-vitro Conservation

Techniques for in vitro conservation of plant germplasm may be applied in any situation where the basics of tissue culture and cell culture are available, and technologies discussed above can be applied to wild species, the crops of agriculture and horticulture and tress. Methods for minimal growth storage are available to most laboratories, and it becomes a matter for local decisions as to whether attempts at conservation should be made and in some way be formalized. This would mean, a commitment to adequate documentation for conservation purpose and to disseminating information about the collection, however limited, to suitable parties.

Cryogenic storage requires relatively simple equipment and facilities, but these are often costly which limits its application. Liquid nitrogen is a relatively common material that can be obtained in most parts of the world and its availability only occasionally restricts the use of cryopreservation in conservation. As this technique provides dramatically extended storage periods, it is perhaps even more important that care is paid to the quality of documentation and record keeping associated with the conservation processes. Major organisations such as IBPGR and ATCC (American Type Culture collection) are in a position to apply, develop and promote in vitro technology in the service of genetic conservation. Importantly, they are sufficiently large and international to be a vehicle for collection, collation and dissemination of information. They have a responsibility to improve in vitro techniques and to be involved in edition whilst such organisations can be seen as an effective world wide model with regard to in vitro conservation. Other national and regional organisations should be prepared to take on a similar rate within their sphere of influence. There are also obvious applications of this technology in all aspects of research and development, including the work of commercial organisations.

Progress in the development of plant cell and tissue culture techniques for long term germplasm. Conservation has been quite significant especially in view of the level of activity and small number of workers involved. The techniques of cell culture in particular have been so refined that only a few species are likely to the recalcitrant to known procedures. During the past decade studies on the freeze preservation of cell culture have trended from basic aspect.

The evidences available so far indicate that freeze preservation is the most reliable approach to the long term preservation of cell culture which possesses the biosynthetic capacity for synthesis and accumulation of secondary metabolites (Kartha, 1987).

Accumulation of secondary metabolites requires a certain degree of differentiation, often accompanied by increase in cellular and volume as well as lengthening of generation period. Therefore the identification of appropriate growth stages most conductive to freezing becomes critical in devising an appropriate cryoperservation strategy of cell producing secondary metabolite.

Germplasm conservation using cell, tissue, organ culture has been reported in 60 species including various crops (Withers 1985, Bajaj 1987). From the information available, conservation is possible in both monocotyledonous and dicotyledonous families. Temperate species figure more prominently then tropical species.

In biotechnology laboratories, freeze preservation and storage of plant material can be of enormous value in maintenance of stock cultures and samples that await screening for future used. Almost an indefinite number of replicates from germplasm of large number of plant species can be stored in vitro using little space in contrast to the thousand of hectares of land that would be required for the same number of plants if maintained in situ. Thus, the material that would otherwise be discarded due to resource can certainly be stored without reservations.

Recent Development In In vitro Conservation

The role of the gaseous environment within the culture vessel in growth and development has been considered with view of extending the relatively low-technology approaches available for in vitro conservation. Volatile compounds in the gaseous culture environment, together with carbon dioxide, will affect growth rates in vitro (Deproff, 1985, Gould and Murashige 1985, Beicher et al., 1987). Other physical factors will be also moderate tc the biological effects of the gaseous environment. The most significant of these to date are the total volume of the gas space above the culture medium and the diffusion distance from the culture tissue to the point of gas exchange with the external atmosphere i.e., the seal of the vessel.

In a study using regenerated carrot plantleis range of culture, vessels were constructed with identical gaseous volumes but differing diffusion distances (Baltson et al., 1988). Biomass measurements at various stages during a 30 day culture cycle showed that growth increase was independent of the gaseous volume above the cultures atleast within the range studies, but was significantly affected by changes in the diffusion distance. Increasing the diffusion distance from 50 to 150 nm, which effectively doubled the period between culture transfers, halved biomass increase. This provides an inexpensive technique that provides some degree of extended storage for in vitro conservation systems.

It is also possible that the advent of gas permeable culture vessels might assist conservation in vitro at the level of unfrozen storage (Tanaka et al., 1988). These containers of various designs are made of relatively inexpensive plastic permeant to water. Their benefits are for organised plantlet cultures in the main, as well as they allow autotropic metabolism to develop to a significant extent, reducing the dependence of the culture upon high levels of exogenous carbohydrates, and similarly reducing the need of complex growth regulators regime. This simplification of the culture environment allows for significantly prolonged growth in culture vessel, with basic nutrients being added using, for example, a hypodermic syringe for sterile transfer of basic medium.

Recent progress in the techniques of induction and manipulation of somatic embryogenesis (Ammirate, 1989) may also have considerable implication of prolonged maintenance of tissue-derived in vitro, with the possibility of a high degree of genetic stability. The technique may be particularly applicable when the material of interest is an isolated cell or protoplast population that can be manipulated into somatic embryogenesis relatively simple and at high frequency. In particular, techniques have been developed that allow for the desiccation of somatic embryos in a way that reflects the physiology and development of the zygotic embryo of the orthodox seed (Ammrito, 1989). The desiccation system relies upon pre-treatment of the somatic embryos with compound such as ABA, IBA and sucrose, added to the culture medium and has been successfully demonstrated for species as diverse as grape (Gray, 1987), soybean (Bendrof and Slawinska, 1988) and wheat (Carman, 1988), whilst not applicable to species whose zygotic embryos are in any case recalcitrant, (Grant, 1986). These developments offer the potential for extended storage of germplasm derived from in vitro system by storing dried, somatic embryos in a manner similar to that used for conventional seed storage. This might provide for long-term storage without an absolute requirement for the relatively advanced technology associated with liquid nitrogen storage systems. If the degree of drying and its associated cellular adaptations do accurately reflect the characteristics of the orthodox seed it might be possible that the material could be kept at room temperature, or the more conventional at 29° C of the mechanical freezer and yet achieve the level of genetic stability required for effective in vitro conservation (Roberts et al., 1981). Further investigation will be required to see if dried somatic embryos can indeed produce the storage characteristics, without genetic alterations of the zygotic orthodox counterparts, coupled with a coating technique used to produce artificial seeds (Redenbaugh et al., 1988), this system has even greater potential.

The relatively recent development of a field collection system where material in the field is taken directly into a protective by possible temporary in vitro environment (Yidana et al., 1987) is a significant enabling technology to be used in conservation. The technique employs various levels of sterilization for explants in the field and incorporates antibiotics into the growth medium. Explants could be collected *in-situ* even in most remote habitats and transported back to the laboratory with the minimum of loss, and without the need for transporting whole plants or large pieces of them.

To date the successful cryopreservation options available for in vitro plants (tissues cultures have largely been based upon slow cooling techniques (Kartha 1985, Withers 1987) although faster rates have been beneficial at times (Grout and Henshow, 1978; Kartha, 1987; Withers et al., 1988). At slow cooling rates there is degree of cellular dehydration following extracellular freezing and then the plunge into liquid nitrogen that typically concludes such slow looking protocols.

To take pure water through the glass transitions temperature to form a glass, as opposed to crystalline ice requires or ulna rapid cooking rates under practical circumstances complete vindication of any significant volume of water is rarely, it ever obtained (Angell, 1982; Reid et al., 1985; Taylor, 1987). A working definition of the vivification does allow for some crystalline ice embedded in a glassy matrix where ice crystal size does not exceed 10 nm diameter (Rienle, 1968). Aqueous solutions such, as those within and surrounding cells in suspension, will vitrify more readily than water alone, and at lower cooking rates, as the solute concentration increases solutions undergo vitrification more readily and so cells incubated with very high levels of cryoprotectant solution can be taken through vitrification as an alternative to the more conventional techniques of cryopreservation. The benefit is that following the appropriate pretreatments required to achieve vitrification, the need for a controlled rate of cooling into liquid nitrogen is removed and simple quenching is likely to be sufficient. Recently, the techniques of vitrification have been successfully applied to Asparagus and Brassica cells cultured in-vitro adding another potential valuable option to the methodology for cryogenic storage (Uragami et. al, 1989; Langis et al., 1989).

In the case of Brassica campestris suspension cells were first equilibrated in 1.5 M ethylene glycol solution and then dehydrated in a complex medium estimated as being approximately 19 osmolal. The high concentration of extracellular solute well result in water loss from the cell, leaving a high concentration of intra-cellular solute also. In this instance vitrification was achieved by plunging a 0.5 ml plastic straw containing cells and medium directly into liquid nitrogen. To avoid crystallization during thawing a rapid warming rates has been employed, which was close to 600' C per min. in this instance. Significant levels of survival were recorded using tripheny tetrazolium chloride (TTC) assay as indicator, and these results confirmed by callus re-growth.

Tissue Culture Approaches

Generally in vitro maintained plant cells, tissues and organs are preserved under appropriate conditions for long terms storage as gene banks. In vitro methods are the most useful for conservation vegetatively propagated plant, species with recalcitrant seeds and genetically engineered materials. The potential advantages of in-vitro conservation include:

1. Requirement for little space for preservation of a large number of clonally multiplied plants.
2. Maintenance of the material in an environment, free of pest or pathogens.
3. Protection against dangers of natural environmental hazards.
4. Availability of nucleus stock to propagate a large number of plants rapidly whenever necessary.
5. Minimizing the obstacles generally imposed by quarantine system on the movement of live plants across national boundaries, since they are raised and maintained in aseptic environment. The present chapter deals only with the application of plant cell and tissue culture in germplasm conservation. Plant tissue culture based biotechnology has opened new avenue for the propagation and improvement of plant species. A wide range of plants has now been regenerated through techniques of tissue culture. Many palm-species including COCK. V nucifera, Oil palm and Phoenix dactylifera (Data palm) multiplies mainly through seeds, and there is no efficient method of vegetative propagation.

The development and application of plant tissue culture system of propagation and conservation would therefore have obvious benefits.

1. Many long lived forest trees including both angiosperm and gymnosperm do not produce seed until a certain age and must be propagated vegetatively when it is desired to reproduce similar to the parental genotype.
2. One area of research currently being investigated is the possibility of very rapid clonal propagation of fruit trees by means of tissue culture techniques. These techniques have special relevance to certain fruit tree cultivars, which are either difficult and/ or expensive to propagate by conventional means.
3. Somatic embryogenesis for higher frequency plant regeneration and synthetic seed production (Gupta et al., 1993).
4. Generating genetic variability for better genetic traits (Larkin and Scowcraft, 1989) somatic hybrid production for increased fields.
5. Cryopreservation of valuable gene pool for future use. (Withers, 1985).
6. Production of cell line and recovery of plants for disease resistance, drought resistance, salinity resistance.
7. In-vitro propagation, improvement and conservation of rare/vulnerable/endangered plant species (Withers, 1986).

In vitro Conservation

Conservation in vitro is wholly dependent upon the techniques of plant cell, tissue and organ culture, and is appropriate in situations where conventional seed storage cannot or is not to be employed.

In particular in-vitro conservation is used for vegetatively propagated material species with recalcitrant seed and material biology. This later material may have been costly and limited in quantity and have characteristics whose stability during repeated mitotic divisions and following meiosis are unknown. It is, therefore prudent to conserve stocks of this material to ensure consistency in subsequent experimentation of production processes, and as definitive source of reference material.

The material stored in vitro may be protoplast, isolated cells grown in suspension or on semi-solid medium, meristem cultures at various stages of development or organised plantlet. It can be assumed that genetic stability within the in-vitro systems increase as the complexity of the cultured material, with completely differentiated plantlets in culture having the least risk of genetic alteration during an in vitro excursion (Karp, 1989). Consequently, if the storage conditions are to permit some level of growth and metabolic activity then organised plantlets are the conservation material of choice, but where growth is to be completely inhabited as in cryopreservation excised shoot their stems may be the more suitable explains (Ford-Lloyd and Jackson 1986; Withers 1987) where cell or protoplast cultures are to be maintained and there is no choice as to a preferred level of organization then care must be taken to ensure the maximum genetic stability, preferable by opting for cryopreservation.

At the simplest level conservation in-vitro is achieved by repeated subculture continuing the protection from adverse environmental conditions and pathogens that is inherent in the invitro systems. This is expensive both in terms of labour and the basic material of the culture process and whilst least demanding in terms of the level of available technology, has a considerable risk of material risk of material loss due to human error or the failure in-vitro security e.g., the invasion of pathogen during transfer.

Modes of Propagation

There are large number of threatened/rare/endangered plant species on which in vitro propagation of threatened plants could be achieved by the following methods :

1. Clonal propagation
2. Somatic embryogenesis
3. Organogenesis
4. Callus differentiation

Clonal Propagation: A variety of plants species can be propagated through forced auxillary shoot bud proliferation. Thus plants produced through this methods are morphologically and genetically similar to the parental plant. Main benefits of these methods are:

1. Rapid multiplication of superior clones and maintenance of uniformity.
2. Multiplication of disease free plants.
3. Multiplication of sexually derived sterile hybrid.

Clonal propagation is achieved by placing sterilized shoot tip or auxillary bud on culture medium that it's sufficient to induce formation of multiple buds.

In vitro propagation of plant using special shod, in nodal segments was obtained in Anogeissus sericea (Kaur et al., 1992). Propagation of Nepenthes khasiana, a threatened plant species endemic to Meghalaya was obtained using nodal segment from the field grown plants (Rathore and Tandon, 1989) and from seedling raising explants (Rathore et al., 1991, Latha and Seent, 1994). Gerbera auranitca, threatened ornamental species recovered by the culturing auxillary shoot buds on nutrient medium (M.S. medium) containing 5 MMBAP, which differentiated into shoots and differentiated shoots include root on IBA (5 to 10 MM) containing medium. A wide range of plants has now been regenerated through this technique.

Somatic Embryogenesis: The process of development of embryogenesis. This was first demonstrated in Carrots (Steward 19. VJ) where bipolar embryos developed from single cell. Further, the tissue derived from these somatic embryos spontaneously developed into embryos again. However, SE is influenced by plant extracts, growth regulator and by the slate of calli.

In vitro propagation of threatened medicinal plants through somatic embryogenesis have been obtained in Podophyllum hexandrum (Arumugam and Bhojwani, 1990), Santalmn album (Bapat et al., 1990).

Mura Costa et al., (1993) have reported in-vitro plant regeneration in Ocoteu ctitharincnxix, an endangered Brazilian hard wood forest plant through somatic embryogenesis.

Research work is also being done to produce artificial seeds via this pathway, but a lot of work has to be done in the area of embryo encapsulation, storage and conversion into plants.

Organogenesis: The process of regeneration of plants from the callus called as organogenesis. In Meconopsis panicutlata, regeneration was observed through organogenesis (Sulaman, 1994) and same type of mode of regeneration has been observed in Meconopsis ximplidfolia, an endangered plant species of Himalayan region (Sulaman et al., 1993). Rheum embodi-a threatened plant was regenerated from shoot tip and leaf explant through organogenesis (Lal and Ahuja 1998).

A number of such examples can be sited which have developed the complete plant through regeneration by organogenesis thus helping in conservation of germplasm.

Callus Differentiation: In vitro regeneration has been reported in Coptis teeta, an endemic and endangered plant of Arunachal Pradesh through callus differentiation into shoots and subsequent induction (Tandon and Rathore 1992). In Valeriana wallichii, a threatened plant was propagated through callus differentiation and recovered complete plantlets (Mathur and Ahuja 1991).

In vitro Storage of Germplasm and Cryopreservation

Assuming that differentiated plantlets are to be stored in-vitro, with all that this implies for genetic stability, and then lowering growth temperature to slow metabolism and development has previously reduced the labour and expense

Table 17.0. In vitro regeneration of threatened/rare/vulnerable/endangered plant species

Name of Plants	*Explant Used*	*Medium*	*Mode of Regeneration*	*Reference*
Aconitum carmichaeli	Shoot tips & Auxillary buds	MS	Clonal multiplication	Hatano et al., 1987
Aconitum nepellus	Nodal shoot	MS	Auxillary shoot proliferation	Robert cervelli, 1981
Aconitum novebora-acense	Nodal shoot	MS	Auxillary shoot proliferation	Robert cervelli, 1981
Anogeissus sericea	Nodal segment (seedling ex-plant)	MS	Auxillary shoot proliferation	Kaur et al., 1992.
Caralluma edu/is	Shoot segments	MS	Clonal multiplication (shoot tip)	Hatanco et al., 1987
Commiphora wightii	shoot segments	MS	Auxillary shoot proliferation	Barve & Metha. 1993
Coptis teeta	Hypocotyl segment	MS	Callus culture	Tandon & Rathore, 1991.
Feronia limonia	auxillary branching	MS	Auxillary shoot proliferation	Purohit & Kiran Tak
Gerbera aurantiaca	Auxillary bud	MS	Auxillary shoot proliferation	Mayer& Vanstaden, 1998
Nepenthes Khasiana	Mature nodal segment	MS	Auxillary shoot proliferation	Rathore & Tandon, 1989.

Contd.

Name of Plants	*Explant Used*	*Medium*	*Mode of Regeneration*	*Reference*
Nepenthes Khasiana	Auxillary bud	MS or Wood Pt. Medium	Auxillary shoot proliferation	Latha & Seeni, 1994.
Ocotea catharinensis	Zygotic embryos	MS	Somatic enbryogenesis	Costs et al., 1993
Oncidium varicosum	Seeding	Knudson medium	Root tip culture	Kerbeay, 1983
Picrorhiza Kurroa	Auxillary bud	MS	Auxillary shoot proliferation	Upadhyay et al., 1989
Podophyllum hexandrum	Zygotic embryos	MS	Somatic embryogenesis	Arumugan & Bhojwani, 1989
Rauvolfia serpentina	Low temperature	MS	Storage	Sharma & Chandel, 1992
Rheum embodi	Auxillary bud	Liquid	Auxillary shoot proliferation	Lal & Ahuja, 1989-93.
Rumex acetosella	Auxillary bud	MS	Somatic embryogenesis	Calafic et al., 1987.
Santalum album	Zygotic embryos	MS	Somatic embryogenesis	Bapal et al.,
Trichopus zeyanicus	Auxillary bud	MS	Auxillary shoot proliferation	Krishan et al., 1995.
Valeriana wallichii	Auxillary bud	MS	Callus culture	Mathur & Ahuja 1991.
Wrightia tomentosa	Auxillary bud	MS	Auxillary shoot proliferation	Purohit et al., 1994.

for repeated subculture. The security of the cultures is also improved as the frequency of the physical interventions of subculture is also reduced. Similarly, the use of somotic stress and the addition of growth retardant to the culture medium have effectively prolonged the interval, between culture transfers (Withers 1987, Pritchard et al., 1986).

The greatest genetic stability is achieved by storage of propagules, capable of regeneration in vitro liquid nitrogen i.e., cryopreservation (Kartha 1985, 1987, Grout and Morris 1987, Withers 1987) where possible these propagules will be excised meristems, capable of direct regeneration into plantlets, but they may, of necessary, by isolated cells or protoplasts.

Cryopreservation

In this system stabiiity is imposed by ultra low temperature and storage is at, or close to-196' C using liquid Nitrogen (or the vapour immediately above it), as practical and convenient oxygen. At such temperature normal cellular chemical reactions do not occur as energy level are too low to allow sufficient molecular motion to complete the reaction. Water exists either in a crystalline or glassy state under these conditions and such high viscosity (> 1013 poises) that, rates of diffusion are insignificant over time spans measured at least as decades. The majority of the chemical changes that might occur in a cell are therefore, effectively prevented and so the cell is stabilized of the maximum extent that is practically possible.

Unfortunately, that is not to say that biological material successfully, cooled to 196' C is in state of complete suspended animation. Certain type of chemical reaction can still occur at these temperatures. Such as the formation of the free radicals and macromolecular damage due to ionizing radiation. The only real threat to genetic stability comes from such reactions, especially those that damage nucleic acids. Any damage that does occur will necessarily be cumulative as enzyme repair mechanism are also totally inhibited at these low temperatures.

While there are as yet few quantitative data on genetic stability at ultra low temperature for higher organisms, studied by Ashwood, Smith and Grant (1977) indicate that to reach a D10 level (Where D10 is the radiation dose resulting in 90% mortality of the population) a frozen cell population would have to be exposed to background radiation for some 32,000 years. It is also noteworthy that dimethyl sulphoxide, probably the most commonly used cryoprotectant in vitro preservation and may aid in radiation damage. (Finkle et al., 1995).

The potential of conservation system for in vitro material based upon cyrogenic storage is therefore, clear and the technique has become relatively widely used.

Other Methods

Cold Storage

Germplasm conservation by storing material in cultures at low and no-freezing temperature the ageing of plant material is slow down but not completely stopped as in freeze prevention.

Low Pressure and Low Oxygen Storage

Attempts have been made to develop Low Pressure Storage (LPS) and Low Oxygen Storage (LOS) as feasible techniques for future conservation of cultured plant materials. In LPS, the atmospheric pressure surrounding the tissue cultures is reduced resulting in partial decrease of the pressure created by gases in contact with the plant material. On the other hand, in LOS the atmospheric pressure (760 mm. Hg) is not reduced but the inert gases (Particularly nitrogen) are combined with oxygen to create low oxygen pressure.

These are found useful in both short-term and long-term storage. There is also belated recognition that nature is infinitely more inventive than chemists and the natural molecules have with stood many tough tests of survival for millions of years. An effort is now onto revive culturing of plant in vitro. A biotechnology firm in USA. Pliytera, has for example set up and stored plant cell cultures of 5,000 spp. in glass vials.

Limitations of In vitro Conservation

There are practical limitation to any conservation system that does not provide a maximum genetic resource whilst, at the same time, conserving the potential of the system to produce whole plants. The level of limitation that is acceptable depends upon individual circumstances. However, where resources are not themselves limiting and the pressure of the time allow for significant research and development, the technique of cyropreservation must be the current in vitro conservation method of choice.

Yet, application of drying to somatic embryos, and the construction of artificial seeds, presents an exciting prospect for the future. Once successfully processed, the dried somatic embryos might be effectively stored at-20' C using a simple domestic refrigerator, which is globally available technology. However, the pre-treatments necessary to achieve desiccation without loss of viability are often complex and currently it is only the tissue of somatic embryos that can be treated in this way.

It is likely that the somatic embryo shares much of its genetic expression and developmental pathways with the zygotic embryo. Both have a potential to withstand extremes of desiccation when the zygotic embryo, and that both have potential to withstand extremes of desiccation when the appropriate sequence of genes are in operation. In a developing orthodox seed (Roberts, 1973) the hydrated, zygotic embryo will respect to chemical stimuli from the parent plant to develop desiccation resistance and thus dry to the low water content characteristic of the seed. It appears that a somatic embryo can be triggered artificially in much the same way (Ammirato, 1989). This technique has obvious potential in circumstances where embryogenesis is both possible and desirable, and where the zygotic embryo does not show recalcitrant characteristics (GROUT 1986). However, it is not clear whether the techniques for drying could be feasible for other somatic cell and tissue types, that have patterns of development and differentiation removed from that of the embryo whilst it is true that leaf callus cell for instance, contain the genes used in the developing embryo to tolerate desiccation during seed formation. It is another matter to be able to exploit this aspect of the genome independently of the rest of the embryo development. (Morris and Grout, 1990).

As cryopreservation has developed as a technology, and surviving material examined more critically, the concepts of optimized procedures and modified culture conditions by support recovery growth following thawing have been considered in more depth.

It is probably an exceptional occurrence when a cryopreserved cell is thawed to physiological temperatures with structure and functions completely unaltered, ready to resume typical activity. More commonly, surviving cells will have varying degree of damage that will be able to repair, depending upon available reserves and metabolic integrity, and may require specific "recovery" media or past thaw conditions to ensure a reliable return to a known and predictable performance (Grout et al., 1976, Benson and Withers 1987, Benson and Narotha Dutra 1988). Those that cannot effect satisfactory repair will die, sometime after thawing.

Different circumstances of cooling and warming may well generate different lesions and consequently differing requirements for repair, and it is not improbable that, even after repair, recover might show a range of alteration in both structure and function.

Cryopreservation of Microorganisms

Reports are available from the studies with microorganisms that suggest suboptimal cryopreservation procedures may cause genetic alternations. Chromosomal damage has been observed in E. coli, apparently caused by freeze thaw stresses during cryopreservation and subsequent recovery with the suggestion that DNA repair mechanisms are important in the recovery from atleast some aspect of freezing stress (Williams and Calcott, 1982). Studies with genetically manipulated yeasts also have shown that viability may persist at high level following cryopreservation.

It is important, therefore, that in vitro conservation be based on cryopreservation procedure that is believed to be optimized.

Present Status of Ex situ Conservation

To complement *in-situ* conservation, attention has been paid to ex-situ conservation measures. According to currently available survey, Central Government and States Governments together run and manage 33 Botanical gardens; Universities have their own Botanic gardens. There are 275 zoos, deer parks, safari parks and aquaria etc. A Central Zoo Authority was set up by Government of India to secure better management of zoos. A scheme entitled assistance to Botanic garden provides onetime assistance to botanic garden to strength and institute measures for ex situ conservation of threatened and endangered species in their respective regions.

Much attention has been paid to genetic conservation of cultivated plants and domesticated animals. Zoological parks have been looked upon essentially as centres of education and recreation. They have played an important role in the conservation of species such as Manipur Thamin Deer and White Winged Wood Duck. Notable examples of successful captive breeding are those of Genetic Gharial, Crocodies, Turtles and white Tiger. NBPGR is playing important role in biodiversity conservation.

2

BIODIVERSITY CONSERVATION: LEGAL ASPECTS

The depletion of biodiversity is an alarming problem all over the world. The World Conservation strategy suggests that the initial efforts of biodiversity conservation should aim at establishment and maintenance of a network of Protected Area (PA) system by making policy changes involving local people in the PA management and mobilizing financial resources for their conservation and protection.

After the 1991 amendments of Wildlife Protection Act, some of the States have been quite vocal about the rigid protection that is being attempted to be given to the protected areas denying the local peoples access to biological resources of this for their use and sustenance.

One has to keep in mind that protected areas make vital contribution to the conservation of the world's natural and cultural resources. Values range from retention of representative samples of natural region, its wild inhabitants and the preservation of biological diversity to the management of environmental stability of the surrounding region viz., the fertility of marginal lands, preservation of a wide variety of genetic resources, maintenance of quality and availability of water, climatic amelioration etc., They also provide scenic and aesthetic surroundings and offer facilities for recreation and tourism.

A well managed forest may be able to provide sustainable supply of a few forest tree or plant species, or a group of species in perpetuity. But there is no model of management yet known which can be applicable within the short span of human life, which can replicate the biodiversity of a given area in totality. It is because of this that the conserved PAs of India attract attention of global community and there is a demand for visits to these areas for the purpose of education and enjoyment of the people. Thus, the range of services and the values that the protected areas offer cannot be compared with other forest areas and therefore, the management objective of protected areas cannot be the same as applicable to other areas in general.

Unfortunately, the PAs all over the developing countries are now subjected to tremendous pressures, especially because of the burgeoning human population followed by use, over use and sometimes abuse of such-areas, leading to destruction of the biological resources in the name sustenance of the local people. In these countries the glare of the Western World and the profligate life style practiced in the developed countries have led to more and more resource use leading to consumerism of higher order. It is an urgent need of the hour to have PAs to preserve the rich biological heritage.

The Commission of National Parks and Protected Areas (CNPPA) of IUCN has been deliberating on this issue since 1978. In the setting Committee meeting of the CNPPA held in 1993, a consensus was reached to have the following set of categories for the PAs. These are given below with the broad management objectives.

Category I	Strict nature Reserve/Wilderness Areas: The protected areas are managed mainly for science or wilderness protection.
Category II	National Park: The protected areas are managed mainly for ecosystem conservation and recreation.

Category III	Natural Monument: The protected areas are managed mainly for conservation of specific natural features.
Category IV	Habitat/Species Management Areas: The protected areas are mainly for conservation through management intervention.
Category V	Protected Landscape/Seascape : The protected areas are managed mainly for landscape/ seascape conservation and recreation.
Category VI	Managed Resource Protected Area: The protected areas are managed mainly for the sustainable use of natural ecosystem.

The joint management of protected areas has to be in the form of eco-development around PAs, through resource sharing to bring out people's participation. This has to be incidental and auxiliary to the management system and not the objective of the system itself.

The National Parks should be kept inviolate so as to maintain biodiversity and continue ecological processes. In sanctuaries, however, some amount of activities are inevitable and there are certain provisions in Wildlife Protection Act to this effect.

Section 29 of the Act says that "no persons shall destroy, exploit or remove any wildlife or its habitat within a sanctuary except under, and in accordance with the permits granted by the Chief Wildlife Warden, and no such permit shall be granted unless the State Government is satisfied that such destruction/ exploitation or removal is necessary for the improvement and better management of wildlife therein". A number of our protected areas are having commercial and pure plantations of Teak, Eucalyptus, Ailanthus, Poplar, Anthocephalus etc., under which, there is hardly any understorey or desirable undergrowth. These are certainly not good as wildlife habitats as they neither provide adequate cover, nor fodder which are essential requirements of wild animals. It would be beneficial to carry out heavy thinning of such plantations and raise desirable species which could provide good fodder and cover to wild animals. This can be feasible in the sanctuaries. The local people living in the fringe areas of such sanctuaries can be employed gainfully in such activities and atleast a part of the produce coming out of such thinning can be distributed to them at concessional rates, or under some mutually agreed terms and conditions. Unfortunately, this is not being visualized in most areas. Similarly there is no bar to take up some activities in open and degraded portion of PAs to develop it as ideal habitat for wild animals.

Very often arguments are put forward that in protected areas no rights of local people can be allowed. This is also not correct as Section 24(2)(c) of the Wildlife Protection Act has a provision to allow, by the collector in consultation with the CWLW, the continuance of any right of any person in or over any land within the limits of the sanctuary. However, obviously the objective is to allow only such rights which are compatible with the long term objectives of conservation of the protected area. The actual situation is that in most of the cases the rights were extinguished at the time of declaration of the reserved forest but were not enforced because the objective of ecosystem conservation or the conservation of biological diversity was not a high priority in the management of our forests earlier. In other places, even if the sanctuaries and national parks were declared, the process of section 19-25 of the WLPA was not followed which extinguishes or admits the rights of local people. It is only when the protected areas were transferred to wildlife wing and serious attempts were made to control the illegal collections or processes under section 19-25 were initiated, the hue and cry blaming that the wildlife laws are creating hardships was raised.

However, it has to be agreed that demarcating out of the protected area system does to some extent, deny general access of the people into such areas and unrestricted use of forest produce in the same pattern as is allowed in other categories of forests.

The traditional life style of the trans-humans was earlier not only ecologically compatible, but also helped in harmonious development of both the indigenous people and the ecosystem. Without entering into the argument whether this was so in all cases earlier, it has to be remembered that in the earlier area forests were plentiful and the population was less. Over centuries, the forest cover has declined and the human population has increased. Thus, today there is a big gap between the demand and the supply capability of the forest/ wildlife. Therefore, if the biomass production from the forest areas or outside is not increased or the alternatives to existing biomass uses is not found out, it will be impossible to meet the demands of the local communities. Apart from that, access to modern living and the profligate style of life has also greatly affected the living style of the people living in or around forest areas which has resulted into increased consumerism and increased pressure of the existing biomass much more than that in past.

It is also advocated that the local communities are gradually getting impatient and alienated to the protection measures. While it is true in some cases, this is not the picture everywhere. In most of the cases, adequate attempts have not been made to educate the local communities about the services that forests and wildlife render towards their sustenance in the long run, and to look into their genuine problems. In absence of such extension measures, some unscrupulous people and opportune seekers take advantage of the ignorance of the illiterate local people and instigate them to resent against any control measure which apparently appears to cut down their access to traditional use of forest and wildlife. On the other hand, where local people have been convinced of the utility of forest and wildlife towards their long term sustenance, the same community which was damaging wildlife and its habitat earlier are now imposing restrictions on the use of habitat and the habitants themselves. Thus, the common people need to be educated about our culture and tradition, and the differences in the scenario that exist today and that of the past, and the reason why the erstwhile-traditional living style, which had not adversely affected the forests till about fifty years back, cannot continue in the present circumstances.

If this is done, a lot of misunderstanding and resentment which is there in a number of areas can be resolved and harmonious growth of people and protected beyond the limit of its sustenance and management objectives.

One has to remember that the forests within reserved and protected forests, as well as protected areas, exist today because of the legal protection offered to such areas. The need of the hour is that the provision of law should be strengthened by change of people's perception towards conservation, combined with certain measures, which will also ensure some amount benefit to the communities and individuals. Some amount of hardships to local people may be caused initially, but these can be partially mitigated through launching of eco-development programmes and fulfilling the biomass needs of the local population. This can be achieved by creating the required forest resources in lands outside the protected areas, and by the introduction of appliances and techniques for efficient, utilisation of energy and biological resources to reduce wastage.

The sense of cooperation within local population would automatically develop for participating with the PA manager not only in the eco-development plan, but also for the entire management planning of the PA.

Joint management in the context of protected areas should essentially cover participation in the protection of all zones and responsible use of resources from the buffer zone of the PAs and multiple-use zones around them. This would amount to improvement in PA management because once the degrading factors are brought under control; the restoration of productivity will be brought about by nature itself.

Legal Instruments Relevant To Biological Diversity In India

Agricultural and Processed Food Products Export Development Authority Act, 1985/1986

- Promotion and regulation of export of agricultural products specified in schedules
- includes medicinal plants

Agricultural Produce (Grading and Marking) Act, 1937

- Fixing grade designations to indicate quality of any specified agricultural produce (3a, b)
- Prohibition or restriction on trade in wrongly marked/graded produce (3g)
- Extention of such provisions to any other article (incl. non-agricultural articles) [6]

Cardamom Act, 1965

- Provisions as in Rubber Act (See below); includes seeds.
- Provision for prohibition/restricting export/import of cardamom (21). Applicable to Elettaria cardamomum Maton, but extendable to any other plant notified by Cardamom Board [3].

Coconut Development Board Act, 1979

- As in Rubber Act, Tea Act, Cardamom Act, etc.

Customs Act, 1962

- Regulation of import-export specifically for:
  (a) The protection of human, animal or plant life or health [11 (t)]
  (b) The conservation of exhaustible natural resources [11 (m)]
- Regulation of transportation and storage of notified items [11 j, k l, m)].

Destructive Insects and Pests Act, 1914,

- Prohibition or regulation of import of any "articles" which may cause infection to any plant [3(1)]
- Prohibition or regulation or movement, between states within India, of articles likely to cause infection to any plant [4A] Note: 'articles' includes insects and plants

Environment (Protection) Act, 1986

- General measures to protect environment [3(i)]

- Restriction of industrial and other processes/activities in specified areas [3(2)(v)] (Read with rule 5 of Environment (Protection), Rules, 1986)
- Prevention and control of hazardous substances, including their manufacture, use, release and movement [3(2), 7, 8]

Fisheries Act, 1897

- Prohibition on use of explosives for fishing [4(1)] Prohibition on use of poisons for fishing [5]
- Regulation on fishing in private waters, with consent of owners/right holders [6(2) & (3)]
- Prohibition on fishing in private waters, with consent of owners/right holders [6(2) & (3)]
- Prohibition of all fishing in specified waters for maximum 2 years [6(4)]

Forest Act, 1927

- Setting up and managing reserved forests [Chapter II] Setting up and managing village forests [Chapter III] Setting up and managing protected forests [Chapter IV]
- Protection on non-government forests and lands [Chapter V]
- Control of movement of forest produce [Chapter VII]
- Control of grazing/trespass by cattle in forest land [Chapter X]

Forest (Conservation) Act, 1980

- Prohibiting or regulating non-forest use of forest lands [2]

Import and Export (Control) Act, 1947

- Prohibition or restriction on imports and exports of specified items [3] Regulations on transportation of specified items [4e]

Marine Products Export Development Authority, 1972

- Establishment of an Authority for developing and controlling marine products [4, 9(1)]
- Developing and regulating off-shore and deep-sea fishing; taking measures for conservation; fixing standards for export, regulating exports [9(2a, c, 0]
- Prohibition/restriction on export and import of marine products [20(1)]

Maritime Zones of India (Regulation of Fishing by foreign vessels) Act, 1980

- Regulation of fishing in India's EEZ by people using foreign vessels [3]
- Permits only to be granted within definition of public interest, and for scientific research, experiments, etc., [5(3), 8]

National Dairy Development Board Act, 1987

- Establishment of a Board which promotes dairy development and other agriculture based industries [4, 16(1a)]
- Financing and facilitating animal husbandry, agriculture, high yielding cattle (including import of semen), import-export of milch animals and bulls and general enhancement of cattle wealth [16(1)]

National Oilseeds and Vegetable Oils Development Board, 1983

- As in Rubber Act, etc., special focus on providing farmers, especially seeds, and certified seeds of high quality, and for improved methods of cultivation.

New Seed Development Policy, 1988

Prevention of Cruelty to Animals Act, 1960

- Restrictions on cruel treatment of animals, including use, transportation, and trade [Chapter III, and Rules under Section 38].
- Restrictions on use of animals for purposes of experimentation and performances [Chapter IV & V].

Rubber (Production and Marketing) Act, 1947

- Establishment of Indian Rubber Board, with function of developing/encouraging improved rubber cultivation and marketing, advising or import/export [BE(1) &. (2)]
- Restrictions on right of rubber planters-license returned to plant or replant where to plant, etc., (171)

Note: Applicable to 4 species of rubber initially, more in Board so notifies [Definitions]

Seeds Act, 1966

- Regulation on quality of seeds of notified food crops, cotton and fodder, to be sold for agricultural purposes [5, 6]
- Restriction on export/import of notified seeds [12]
- Exemption to persons selling/delivering on own premises, seeds grown by them [24]

Species Board Act, 1986

As in Rubber Act, etc., for cardamom; for other spices, restricted to exportimport development and regulation.

Tea Act, 1935

- As in Rubber Act, etc., includes restrictions on export of tea seeds [17].
- Applicable to one species, Camellia sinensis, presumably to all the varieties.

Territorial Waters, Continental Shelf, Exclusive Economic Zone and other Maritime Zones Act, 1976

- Establishment of sovereign rights over waters and seabed within the continental shelf and exclusive economic zone (200 nautical miles from nearest appropriate point on Indian territory) [3(1), 5(1), 6(2), 7(4)].
- Sovereign right to explore, exploit, conserve and manage resources of continental shelf and EEZ [6(3),7(4)]

- Notification of any area within this zone for purpose of protection of resources and conservation of marine environment [6(5),7(6)]

Tobacco Board Act, 1975

- As in Rubber Act, Tea Act, Cardamom Act, etc.

Wildlife (Protection) Act, 1972 and Wildlife (Protection) Amendment Act, 1991

- Restriction or prohibition on hunting of animals [Chapter III]Protection of specified plants [Chapter IIIA]
- Setting up and managing sanctuaries and national parks [Chapter IV]
- Setting up of zoo authority, control of zoo, : adoptive breeding [Chapter IV A]
- Control of trade and commerce in wild animals, animal articles and trophies [Chapter V and Chapter VA]

Biodiversity: Policy and Priority Setting

Out of the several international mechanisms and bodies of the United Nations actively involved in the area of conservation and sustainable and equitable use of agro-biodiversity/plant genetic resources the FAO's intergovernmental Commission on Plant Genetic Resources has been playing the leading role globally. The FAO Global System provided a non-binding International undertaking on Plant Genetic Resources (IUPGR) since 1983, which advocated that these resources should be made freely available, being heritage of mankind. In a revised agreed interpretation it stated that the free access does not mean free of charge. The contribution of the FAO global system in the PGR activities has been tremendous and highly esteemed by countries at large even though it did not have a legal entity. On the other hand, the Convention on Biological Diversity (CBD), which is only instrument of international law on this subject, since 1993, reaffirmed building upon the principle that the countries have sovereign rights over their biological resources. The CBD also aimed at conservation of biodiversity, sustainable use of its components and fair and equitable sharing of benefits arising from inter alia use of plant genetic resources. The realization of the Breeders' Rights (BR), the farmers rights (FR) or the issue of access on mutually agreed terms to the plant genetic resources, including ex situ base collections of CGIAR centres not covered by the CBD, provided the basis of a strong interface between these two institution/mechanisms and a revision of the FAO's international Undertaking accordingly being debated to bring harmony between the two.

The basic principle that nations have sovereign rights over their genetic resources was stated earlier by the UN Conference on Human environment in 1972 at Stockholm and enshrined subsequently by the CBD. There are, however, technical limitations for exclusiveness or control ability of access to genetic resources which hamper compliance with laws regulating the access. The instruments of mutually agreed terms and prior informed consent (PIC) have been brought in collecting and Transfer (ICCPGCT) of the FAO Global System. The issue of approval and involvement of holders of traditional knowledge for the sustainable use of biological diversity is considered to have wider application and has been duly addressed by the CBD (article 8). However, CBD does not specifically outline as to how the domestic PIC (of local communities and farmers) should be obtained and organized.

The Trade Related intellectual Property Rights System (TRIPS) of the World Trade Organization (WTO) provide for another legal dimension. According to this, the nations are required to evolve a sui generis system for varietal protection, go in for patenting or adopt a combination of both, but not in isolation from the broader goals set out in agenda 21 of the CBD, particularly chapter 8. A sui generis system, in its absolute sense, provides an institutional mechanism to manage and enrich the biological diversity, which includes the agro-biodiversity, within the national territory, giving due opportunities to its safe conservation and use besides protecting and safeguarding the interests of those who conserve and maintain its vital components onfarm. Issues that nations are required to consider, plan for and implement in an effective legal and/ or regulatory frame work, include the following:

1. Regulation of exchange/flow of germplasm in conformity with the charter of the UN and the international law.
2. Use of sovereign resources and agro-biodiversity not to have any harmful effects on territory beyond national borders.
3. Pipeline protection to plants or plant parts patented elsewhere, etc.

The agreed scientific principles and practices for on-farm genetic resource conservation are lacking. Usable classification and nomenclature systems for landraces, as well as methodology for their characterization is also warranted which have to be duly studied side by side while implementing the on-farm conservation programmes. The temporal and spatial movement of genes and varieties into and out of cropping systems has to be better understood as also the causes of loss or rejection of some varieties or traits over others. Some of the traditional practices have to be adequately studied and indexed. Also, methods to maintain healthy populations of beneficial insects and soil microorganisms have to be worked out.

The CBD expects parties not to practice national decisions in isolation from the broader goals of sustainable development set out in Agenda 21. Thus, emphasis has to be given to integrate economic, social and environmental factors at the policy, planning and management levels, supported by effective legal and regulatory framework and also by the appropriate provision of incentives. Integrated environmental and economic accounting is important particularly for implementation of sustainable development programmes. Role and relationship of private sector on issues of agricultural biological diversity is highly voiced, which needs to be properly defined, elaborated and legally framed. The CBD by and large, also provides basis for the access to genetic resources vis-a-vis technology. Working groups have yet to explore the prospects of promoting international initiatives in respect of Articles 16(1) and 18(2) on biotechnology and scientific and technical cooperation, respectively.

India has been committed to international cooperation on mutually agreed terms with parties, being the signatory to the International Undertaking on free exchange of plant genetic resources. The areas of agro-biodiversity conservation and capacity building are well focussed in the Global Plan of Action, adopted during IV Technical Conference at Leipzing during 1996. Whereas the ex situ conservation is adequately being addressed by several national, regional and international all gene banks, the in situ on-farm conservation still needs a lot or preparatory exercise in planing, developing policy and farming elaborate plan of action. Capacity building for on farm conservation of traditional "types" requires formulation of suitable policies, sector and cross sector plans and programmes for implementation and also interim finical mechanisms. Recognition of the associated local knowledge, informal innovations and practices are needed to promote traditional farming systems, whereas establishing community seed banks which are ultimately linked to the base collection at the national gene bank which are ultimately linked to the base collection at the national gene bank, developing and maintaining germplasm/landrace registration system, etc., is required for a smooth implementation of IPR. The recognition of and formulating implementable laws and regulations for "Farmers Rights" is equally important.

The intra-specific variation and the inter-dependence of each variety on its agro-ecology and socio-economic setting continue important areas of research in these modern agro-ecosystem. The traditional breeding programmes have to be supplemented with molecular DNA characterization laboratories for refinement of research or verification technology since the diversity within the genetic make put of a crop is more pertinent than the variation for morphoagronomical characters alone.

Improving and promoting use of marginal or underutilized crops needs prioritization beyond their "on-farm" conservation activates. This would help in crop diversification as well as improvement in agro-ecology. Other research priorities include.

1. Mass multiplication of endangered or disproportioned landraces by use of biotechnology.
2. Genetic characterization of strains of micro-organisms likely to be used as inoculates for fertility (Nitrogen level) restoration of lands degraded under intensive cropping, and.
3. Documentation of case Indies of conservative and low external input production systems.
4. Marking provision for incentives and rewards to innovators developing readily transferable eco-friendly and sound technologies in relation to conservation and sustainable use of agro-biodiversity. At the same time, the risks associated with the introduction of alien species or genetically modified organisms into the modern commercial agriculture have generated concern of late. Therefore, capacity building for bio-safety and eradication of threatening alien species, through legislation, may have to be reviewed seriously.

Biodiversity : Institutional Mechanism And Modalities

The Institutional mechanism at the national level is provided by the Ministry of Agriculture in conjunction with the other Ministries, such as, law, Commerce, Finance, Forestry and Environment, etc., The ICARDARE, as the nodal organization, spearheads, the planning, coordination and implementation of all activities on agro-biodiversity in the public sector and maintains linkages with other Government Departments, Communities, NGOs and private sector. Factors such as, environment, cropping system, socio-economics have been integrated in the system through appropriately organizing a balanced system of conservation, sustainable development and progressive farming.

The ICAR has been providing a major funding support to State Agricultural Universities (SAUs) and also carrying out research, research management, research co-ordination, training and agricultural education activities as its own institutional set-up to bridge the gaps. It is an old and experienced organization and is a leading National Agricultural Research System (NARS) globally. The national needs of agricultural research, extension and education have been diligently and meaningfully catered by the Council all the while, despite the fact that agriculture is a State Subject as per the Constitution of India. There is a further need to self-determine and fulfil such responsibilities for the system, giving due importance, opportunities and avenues to emerging new linkages.

The medium to long term perspective planning of the ICAR amplifies Council's concern for achieving a combination of sustainability and advancement in the field of agricultural research, including conservation and rational use of agro-biodiversity. The ICAR provides funds and manpower support to several university based centres to work on underutilized crops realizing scientific need to develop technology for new crops of future potential. Also, in order to strengthen research and advance technology base for rapid characterization for diversity in plant samples, a National Research Centre (NRC) on DNA finger printing has been established, besides and NRC on plant biotechnology.

The India's National Plant Genetic Resources System

The India National Plant Genetic Resources System (IN-PGRS), under the aegis of the Indian Council of Agricultural Research (ICAR) and spearheaded by the National Bureau of Plant Genetic Resources (NBPGR) is among the most dynamic systems in the world which now hold a prominent place and a leading gene-bank. The NBPGR has been entrusted with the national responsibility to plan, conduit, promote, co-ordinate and take lead in activities concerning germplasm collection, introduction, exchange, evaluation, documentation, conservation and sustainable management of diverse germplasm of crop plants and their wild relatives with a view to ensure their availability for use over time to breeders and other researchers. It includes NBPGR network of 10 regional stations/base centres/quarantine centres over different phyto-geographic zones of the country and active collaboration and linkages with over 30 National Active germplasm Sites (NAGS).

DARE/ICAR is the umbrella organization while NBPGR is the nodal agency. It controls 11 Regional Stations located all across the country as per agro-ecological analogues. Apart from it NBPGR also have National Active Germplasm Sites (NAGS), ICAR insatiate/project, Directorates/National Research Centres/All India Coordinated Research Projects/ State Agricultural Universities, Other National Stake Holder and have International Collaboration.

The INPGRS is responsible for conservation and management of PGR and their use through an effective collaboration between the NBPGR and the user agencies. NBPGR is the nodal organization operating the Indian PGR system and has the authority to import and export of plant germplasm for research purpose through the single window system. It is also fully equipped with very effective and efficient plant quarantine facilities. NBPGR has assisted over a dozen collaborating institutes/centres in establishing medium-term seed storage; computer and data documentation faculties besides imparting need-based, on job training to their PGR personnel.

NBPGR is helping the collaborating institutions in developing medium term seed storage faculties and also computer faculties for documentation. Bureau also provides training to their scientist and technicians.

The emerging role of non-governmental organizations (NGOs) in managing agro-biodiversity for its conservation and sustainable and equitable use through imparting awareness to indigenous communities and farming and families and developing indigenous knowledge base banks is being widely recognized. This adds to the responsibility of the ICAR, being nodal organization in the ICAR, being nodal organization in the country's agricultural research system, to perceive, anticipate, plan, and frame appropriate policy issues and also to develop model/pilot scale projects related to agro-biodiversity management, conservation and sustainable use. Issue of compensation to farming communities may also be addressed as and where applicable. This, obviously calls for concerted efforts towards understanding the following components:

Classification, indexing and inventorization of traditional agricultural systems, still in vogue, with respect to their eco-systems, crop combinations, seasons any/or system of cultivation.

The ICAR's new initiatives in perspective planning include Agricultural Research Information System (ARIS), Agricultural Human Resource Development (AHRD), National Agricultural Technology Project (NATP) for technology generation and transfer, Institution Village Linkage Programme (IVLP), interface with Ministerial Departments/Ministries of Govt. of India, such as, Do AC DAH&D, DBT, CSIR, DRD, C APART, MoEF, etc., and establishing partnership with NGO and private sectors. Besides involvement in several regional fora and programmes through bilateral agreements and MOUs the Council is posed well to play major role in establishing the country's lead for agricultural development and management of its agro-biodiversity.

Biodiversity Conservation: Future Strategies for India

If you plan for one year, plant rice,

if you plan for ten years, plant trees.

But, if you plan for one hundred years, educate the people.

Development, both individual and societal, and environment are two sides of the same coin. Due recognition is needed for global environmental risks while designing any environmental strategy. Environmentally and socially descriptive economic behaviour must be changed if we intend to provide discarding environmentally insensitive forms of development can only bring about decent livelihood to all citizens of the present and future generations and this change, rather, the stress should be on qualitative development based on efficient use of energy and natural resources, recycling of end products and minimizing the use of non-biodegradable products.

In a country with a burgeoning human and livestock population, concrete actions and sound strategies are easy to formulate but difficult to implement. Human beings have a tendency to prefer present goods to future goods and in the process, overlook the side effects. To sustain 1.5 billion people in India by the year 2025 heavy financial investment and state-of-the-art technology will be needed in agriculture, water resources, infrastructural development and so forth. Ecological balance will definitely be adversely affected to create liquidity despite our best efforts not to do so. Primary and secondary forest industries will use more and more wood, cork, bamboo, cane, gum, resins, essential oils, seeds, medicines etc., for home consumption as well as export in a liberalized economy.

Despite all adversities one has to adopt viable strategies for the conservation of species as this will only guarantee our future and lead us towards the goal of sustainable society, sustainable economy, sustainable use and sustainable development. The strategy should be aimed at policy makers, all Government and Non-Governmental organisations, private organisations, individuals and inter-governmental organisations. This will help in achieving the objectives of biodiversity conservation and sustainable development. Any conservation measure has to be carefully planned so that:

1. Environmentally and socially disruptive behaviour is altered to provide a decent living for all.
2. Production pattern is socially equitable
3. A human centered, knowledge intensive civilization is built.
4. A balance is maintained between all forms of capital i.e., human, natural, physical and financial.

Some of the measures suggested as a part of long term planning are as follows:

Carrying Capacity Of The System

Carrying capacity varies from region to region depending upon population pressure, their dependence on food, water, energy, raw material requirements, waste production and import and export. In order to live within the carrying capacity, the society will have to change the life style and consumption pattern in such a manner that causes least damage to the ecosystem. Over population in a particular region has to be avoided and excessive pressure on land must be reduced. It calls for changes in the present land use policy, overhauling of land tenurial system and population control. Forest land will have to be used in such a way that maintains their biodiversity, productivity, regeneration capacity, vitality and their potential to fulfil, now and in future, relevant ecological, economic and social functions.

Enhancement Of The Protected Areas

One of the best means of protecting the natural forest is through the establishment of protected areas, an effort that has taken more than a century to evolve from the concept of reserve forest to national parks and biosphere reserves. These protected areas serve three basic functions viz., benefit, education and recreation of the people. The area under protected area cover is follows:

Year	*Area (Sq. kms)*	*No. of National Parks and Sanctuaries*
1970	25,00010	NP and 127 Sane.
1987	1,10,00054	NP and 372 Sane.
1992	1,32,00066	NP and 421 Sane.

Although, India has made steady progress in the management and expansion of the protected areas, yet, the problems are being compounded with time. Mere establishment of protected areas does not ensure socio-economic development and species conservation. The management has to take timely action for the speedy disposal of the settlement problems. Local people who consider themselves as the primary stakeholders must be involved in the use and management of the resources. Socio-economic well being of the people should be one of the top priorities for minimizing the resource dependency on the core of these protected areas. Biological resources in the protected areas can be assigned consumptive and non-consumptive values so that the conservation of biological diversity can be linked to sustainable socio-economic development.

At the same time, there is a need for change in the attitude of the field staff towards the interests of the local people. This conflictual relationship has to be transformed into cooperation and partnership. Field staff will have to appreciate that with inadequate workforce prohibitive management strategy is impossible to implement. Therefore, in the long run a conciliatory, interactive and motivational approach seems to be the best.

Enhancement of protected areas will need effective management with clear objectives for each area. This will require increased investment in human resources, education and research.

Stress Of Integrated Development

Of late it has been realized that a direct stress on conservation measures and neglecting the economic development of the people in and around the protected areas leads to impoverishment, and initiates the vicious cycle of subsistence, emigration, land fragmentation, degradation and deforestation. According to the new concept, people will have to change their consumption pattern and find alternatives on their own, once their purchasing power increases. The only fear in this integrated effort is that short-term conflicts may arise in the absence of appropriate policy measures causing hardships to the people. It is therefore essential that conservation, rural development and compensatory incentive programmes should be organised simultaneously.

Research Impetus

There are five main areas, which need urgent attention:

Forest management

Forest management has undergone rapid changes in the decades of eighties and the nineties. Geographical Information System and Remote Sensing Techniques have the potential to provide a large amount of information on forest resources, growth and biodiversity.

Restoration of degraded lands

Research efforts should concentrate on enhancing the soil micro fauna and fertility status so that sustainable production can be achieved by adopting agro-forestry systems that can support the economic development of the region. Efforts should be made for genetic improvement and dissemination of multipurpose tree species that can grow in degraded areas.

Research in harvesting, processing and marketing

There is a need to develop low impact harvesting systems, drying and conversion processes for non-commercial abundant species, especially the NTFP's. Marketing management has been one of the weakest sectors of forestry and needs new directions and support.

Policy, institutional and socio-economic research

Socio-economic objectives are best pursued if the policies regarding land use, rehabilitation of degraded lands, benefit sharing, land and tree tenure, participatory planning and resource management are appropriate to the sustainable management and biodiversity conservation.

Research in biotechnology

Sustainable development and biodiversity conservation in the 21st century will depend upon proper exploitation of economic potential of genetic diversity using biotechnology for the production of goods and services required by the increasing population. Foresters will have a dual role in future; one as in situ conservationists and the other in ex situ maintenance of biodiversity for commercial use.

In addition, social and anthropological research will also be required to determine individual and community needs and perspectives of forests, trees and environment. This will have immense impact on the participatory planning and management processes.

Last but not least the concept of conservation and development should permeate our thinking and perception. We are all stakeholders, participants and decision makers in this struggle and can achieve the objectives by our concrete actions, strength and determination with the active support and participation of the people.

Strategies, Opportunities And Options

In India, creation of Sanctuaries, National Parks and Biosphere Reserves has been adopted as the major task of achieving the objectives of conservation. Restricting human interference to conserve biological diversity is an act opposed by the rural tribals already facing complex problems, including small land holdings, meager employment opportunities and, over and above, natural hazards, reduced productivity of jhunt fields and environmental uncertainties. Conservation strategies must address to the needs of posterity but, at the same time, should not neglect the present needs of the people.

Protected areas in the context of available forests or geographical area of the particular region could be a debatable question. There should be, however, a critical review of the impact of PAs and biodiversity conservation efforts on rural population or indigenous people of the region. The evaluation should cover all the aspects of economic and employment opportunities, cultural diversity crisis, depletion in indigenous knowledge systems and many other similar parameters.

Biodiversity conservation strategies and goals, should redefine biodiversity in the context of local needs, usages and values as priority of conservation. The current experience in the region is that there is lack of objectives of biodiversity conservation strategy, which need to be explicit, clear and achievable. It should not neglect the anthropocentric objective addressing the local peoples' priorities to provide, if not enhance, sustainable and economic benefits.

Current policies and practices of exclusive protectionism of conservation areas need a re-evaluation. After all, people and their culture have historically played significant role in conservation and management of what was there in the past and what is remaining now. There is the need to survey peoples' needs and attitudes around our conservation areas. Possibilities of controlled harvesting or sustainable use of at least certain wildlife resources and minor forest produces (MFPs) by the communities neighbouring protected areas should be considered. Activities that may be taken up as pilot projects may be identified. Such long-term goal may change peoples' attitude towards PAs positively and win their support for wildlife and biodiversity conservation. Such objectives may also ensure long-term survival of wildlife and biodiversity conservation, since the people themselves would want to protect the protected areas. On the other hand, damage to property and human life by wildlife, and the amount and system of compensation payments are some of the issues that are not yet well documented.

Many of the protected areas in India are indeed forests for wildlife shelter in the midst of intense human pressure and activities. Such islands of shelter cannot always afford to accommodate many of the wild animals who possess migratory instinct. There is, thus, the urgent need of identifying animal corridors and bringing such corridors within buffer zones or some kind of management practices in order to reduce man-animal conflicts or wildlife depredation.

In India, it is a fact that there exists rich biodiversity even outside the PAs, such as in sacred groves, traditional village/community/individual forests, in agricultural jhum fields and fallow lands. Such Lands/Forests are under the ownership/control of the Village Councils, District Councils, traditional chief's or private individuals. These traditional socio-political institutions must be taken into confidence while formulating strategies for biodiversity conservation. Management strategies and policies should ideally encompass the areas under the control of these institutions, which can be achieved through appropriate dialogue and negotiations without infringing on their land rights/ownership.

Poverty, coupled with increasing marginalisation of communities. living around the protected areas and increasing number of players/interest groups over limited resources are some of the root causes of people-PA conflict in this region. Unless financial income is generated and there is economic slake for the people for protecting these areas, the problems could multiply and biodiversity conservation goals may not be achieved to the desired level. There is ample scope for amalgamation of traditional practices and modern scientific knowledge for solving the problems of PAs and biodiversity conservation in the region. Not that all traditional societies are always wise stewards of biodiversity or their practices are scientific, conservation-oriented or sustainable. Such practices should also be identified and discouraged and alternatives evolved.

One of the weaknesses of current conservation strategy in the region is its heavy emphasis on conservation of wilderness areas of forest lands. Conservation of cultivated crops, ethnobiological species including medicinal and wild edible plants is a neglected frontier. This weakness must be rectified immediately as the region has a variety of food and medicinal plants, many of which are not much known outside the region and may have potential of providing alternatives for food and medicines in the region and elsewhere. Such crops have also been reported to be pest resistant and stand long storage.

In various silvicultural practices and compensatory plantations in the region, there is often a tendency to patronize one or more better performing species such as pine in the higher altitude and teak in the lower elevations. Such practices should be reviewed and minimised, as monocultures reduce the spectrum of biodiversity, both of plants and animals. Ideally, there should be an assessment of areas that may be covered under monocultures and native mixed species should be planted to maintain biological diversity.

The development projects in India are not sensitive to biodiversity issues. Many of them are harmful to biodiversity to a greater or lesser extent. Therefore, there is a necessity of improvement in project appraisal methodologies and implementation of programmes with a view to making them more sensitive to biodiversity, while at the same time, ensuring sustainable development, better valuation of renewable resources, improved forests/wetlands/wildlands conservation and increased agricultural productivity. There is also the need for sensitizing the project authorities towards biodiversities awareness and initiatives.

In India, the Environmental Impact Assessments (EIA) in any region tends to concentrate on local physical, biological and ecological impacts, their effect on the human population and local environment. However, their effect on the biodiversity is invariably neglected. One of the main practical constraints to improve EIA is lack of data and the difficulty of ascribing values to all but direct use-values of biodiversity.

Inter departmental approach

There is also a lack of inter departmental approach and vision in relation to biodiversity conservation in the region. For example, while the State Environment & Forests/Wildlife Department may be enthusiastic to conserve wetlands, the Fishery Department may like to promote commercial fish farming in such wetlands, or the Agriculture Department may like to reclaim the swamp areas for wet rich cultivation Moreover, there is often random introduction of exotic species or hybrid varieties for short term commercial gain endangering patronage of local species or varieties. Such practices also dilute the indigenous knowledge systems or erode germplasm of the native species.

Urgent Need

There is an urgent need for an increase in research activities, information processing and dissemination system, either to improve priority setting, or to guide management of biodiversity conservation, both within and outside the protected areas and even for almost all habitat types. Funds and resources may be allocated to the best possible effect. There is also the need for chronicling traditional natural resource management systems, which should be evaluated, strengthened and encouraged. Such knowledge should be exchanged or transformed between tribes for higher ecological and economic efficiencies, particularly conservation and management of biodiversity under their control.

Environmental Education

Environmental education, awareness building and motivation initiatives on biodiversity issues using local arts, cultures and languages should also be part of the strategy. The traditional plant and wildlife knowledge and skills should be identified, documented and used as educational tools. There is also the need to infuse traditional values, culture, skills and knowledge into the young generation, many of whom are now exposed to the mainstream life and culture, for better appreciation of their own native wisdom and resources. Environmental education, particularly awareness in relation to implication of biodiversity trade, conventions regulations, removal and extraction of endangered species, etc., should also be extended to the law enforcing agencies.

Future Strategy For The Conservation Of Biological Diversity

The rate of growth of human population must be reduced. Greater emphasis should be given to keen management of available biological resources rather than to just consider conservation of biological diversity in isolation. The reason is that increase production of fuel wood, fodder, building material etc., in areas near in biodiversity in the country is essentially required to onset the pressure on biodiversity rich areas. A natural resource policy may be formulated to achieve four basic goals:

1. Maintenances of adequate resources
2. Conservation of resources through reduction achievement of greater end use in demand and

3. Maximum use of renewable resources
4. Reduction in dependency of non-renewable resources.

Unplanned development works and over-exploitation of resources has not only resulted in shortages of various materials but also left our biodiversity exposed to various ecological threats. The world is rapidly running out of all major non-renewable minerals that are necessary for an industrialized society. Many experts predict that the various economies of the earth will exhaust about half of the world's useful metals in 75 years or less (Henning and Mangun, 1989). Attempts must also be made to rehabilitate degraded lands so as to increase yields in terms of agricultural, forestry, industrial and medicinal products on human dominated lands, and to reduce pressure on remaining natural habitats. It has been shown that about 43% of world's terrestrial vegetated surface has diminished capacity to provide benefits to humanity because of direct impacts of land use (Daily, 1995). Therefore, sound management of biological as well as non-biological resources is essential for the overall preservation of biological diversity.

The agricultural policy should be revised to incorporate conservation of indigenous crop and livestock varieties, perhaps by redirecting subsidies and incentives towards them. A separate overreaching legislation on biodiversity should be considered. The laws on the biodiversity conservation must be strengthened and biodiversity concerns should be reflected in the existing Environmental Impact Assessment procedures. Current conservation programme should be reviewed. The conservation programme in India has focussed mainly on some glamorous animals (mostly large mammals and birds) while ignoring the rest of biological diversity including a host of insects and invertebrate species and a majority of the plant species. Conservation efforts should involve local communities by guaranteeing their land and forest rights and providing them a stake in the sustainability of their natural surrounds.

Some important recommendations have been made by the Core Committee of Ministry of Environment & Forests, which include careful sustainable management and utilization of fossil fuels, high priority to efforts in improving energy use efficiency and conservation, and increased emphasis to research and development of solar techniques and building up of biomass energy sources. If the rate of biodiversity loss is to be slowed down, a combined procedure, project and policy approach would be required. Procedural action would include giving more weightage to biodiversity conservation in project and programme design and appraisal. The project approach could provide immediate and targeted action on priority sites. Policy actions would include appropriate governmental intervention in various sectors through reforms and procedures to promote ecologically prudent behaviour on a broad front.

Biodiversity: Institutional Mechanism and Modalities

The Institutional mechanism at the national level is provided by the Ministry of Agriculture in conjunction with the other Ministries, such as, Law, Commerce, Finance, Forestry and Environment, etc., The ICAR-DARE, as the nodal organization, spearheads, the planning, coordination and implementation of all activities on agro-biodiversity in the public sector and maintains linkages with other Government Departments, Communities, NGOs and private sector. Factors such as, environment, cropping system, socio-economics have been integrated in the system through appropriately organizing a balanced system of conservation, sustainable development and progressive farming.

The ICAR has been providing a major funding support to state agricultural universities (SAUs) and also carrying out research, research management, research co-ordination, training and agricultural education activities as its own institutional set-up to bride the gaps. It is an old and experienced organizational and is a leading National Agricultural Research System (NARS) globally. The national needs of agricultural research, extension and education have been diligently and meaningfully catered by the Council all the while, despite the fact that 'agriculture' is a "State Subject' as per the Constitution of India. There is a further need to self-determine and fulfill such responsibilities for the system, giving due importance, opportunities and avenues to emerging new linkages.

The ICAR has been providing a major funding support to State Agricultural Universities (SAUs) and also carrying out research management, research coordination, training and agricultural education activities as its own institutional set-up to bridge the gaps. It is an old and experienced organization and is a leading National Agricultural Research System (NARS) globally. The national needs of agricultural research, extension and education have been diligently and meaningfully catered-by the Council all the while, despite the fact that 'agriculture' is a "State Subject' as per the Constitution of India. There is a further need to self-determine and fulfill such responsiblities for the system, giving due importance, opportunities and avenues to emerging new linkages.

The medium to long term perspective planning of the ICAR amplifies Council's concern for achieving a combination of sustainability and advancement in the field of agricultural research, including conservation and rational use of agro-biodiversity. The ICAR provides funds and manpower support to several university based centres to work on underutilized crops realizing scientific need to develop technology for new crops of future potential. Also, in order to strengthen research and advance technology base for rapid characterization for diversity in plant samples, a National Research Centre (NRC) on DNA finger printing has been established, besides and NRC on plant biotechnology.

The India National Plant Genetic Resources System (IN-PGRS)

The India National Plant Genetic Resources System (IN-PGRS). under the aegis of the Indian Council of Agricultural Research (ICAR) and spearheaded by the National Bureau of

Plant Genetic Resources (NBPGR) is among the most dynamic systems in the world which now hold a prominent place and a leading gene-bank. The NBPGR has been entrusted with the national responsibility to plan, conduit, promote, co-ordinate and take lead in activities concerning germplasm collection, introduction, exchange, evaluation, documentation, conservation and sustain able management of diverse germplasm of crop plants and their wild relatives with a view to ensure their availability for use over time to breeders and other researchers. It includes NBPGR network of 10 regional stations/base centres/quarantine centres over different phyto-geographic/ones of the country and active collaboration and linkages with over 30 National Active Germplasm Sites (NAGS.

DARE/ICAR is the umbrella organization while NRPGR is the nodal agency. It controls 11 Regional Stations located all across the country as per agro-ecological analogues. Apart from it NBPGR also have National Active Germplasm Sites (NAGS), ICAR insatiate/project, Directorates/National Research Centres/All India Coordinated Research Projects/ State Agricultural Universities, Other National Stake Holder and have International Collaboration.

The IN-PGRS is responsible for conservation and management of PGR and their use through an effective collaboration between the NBPGR and the user agencies. NBPGR is the nodal organization operating the Indian PGR system and has the authority to import and export of plant germplasm for research purpose through the single window system. It is also fully equipped with very effective and efficient plant quarantine facilities. NBPGR has assisted over a dozen collaborating institutes/centres in establishing medium-term seed storage; computer and data documentation faculties besides imparting need-based, on job training to their PGR personnel.

NBPGR is helping the collaborating institutions in developing medium term seed storage faculties and also computer faculties for documentation. Bureau also provides training to their scientist and technicians.

The emerging role of non-governmental organizations CNGOs) in managing agro-biodiversity for its conservation and sustainable and equitable use, through imparting awareness to indigenous communities and larminu and families and developing indigenous knowledge base banks is being widely recognized. This adds to the responsibility of the ICAR. being nodal organization in the ICAR, being nodal organization in the country's agricultural research system, to perceive, anticipate, plan, and frame appropriate policy issues and also to develop model/pilot scale projects related to agro-biodiversity management, conservation and sustainable use. Issue of compensation to farming communities may also be addressed as and there applicable. This, obviously calls for concerted efforts towards understanding the following components:

Classification, indexing and inventorization of traditional agricultural sytems, still in vogue, with respect to their eco-systems, crop combinations. casons any/or system of cultivation.

PGR Activities In Wild Genepool

The National Bureau of Plant Genetic Resources (NBPGR) has been collecting wild relatives of crop plants. General and specific (Ahetnioschm. wlanuni, temperate fruits, Cucurbits) explorations have been undertaken. Inventorization of species wealth available in wild relatives of crop plant, has been published which is being updated, giving a broader and wider perspective on its distribution and extent of diversity, and conservation and use of this diversity. The information points out to much less utilization of wild species while these have great direct and indirect potential in crop improvement, and even as new domesticates.

The ICAR's new initiatives in perspective planning include Agricultural Research Information System (ARIS), Agricultural Human Resource Development (AHRD), National Agricultural

Technology Project (NATP) for technology generation and transfer, institution village linkage programme (IVLP), interface with Ministerial Departments/Ministries of Govt. of India, such as, DoAC DAH&D, DBT, CSIR, DRD, CAPART, MoEF, etc., and establishing partnership with NGO and private sectors. Besides involvement in several regional fora and programmes through bilateral agreements and MOUs, the Council is posed well to play major role in establishing the country's lead for agricultural development and management of its agro-biodiversity.

3

METHODS USED BIODIVERSITY CONSERVATION

The man lived in a hunter-gather society for much of the time and thus depended entirely on biodiversity for sustenance. But, with the increased dependence on agriculture and industrialisation, the emphasis on biodiversity has decreased. Indeed, the biodiversity, in wild and domesticated forms, is the source for most of humanity food, medicine, clothing and housing, much of the cultural diversity and most of the intellectual and spiritual inspiration. It is, without doubt, the very basis of life. Further that, a quarter of the earth's total biological diversity amounting to a million species, which might be useful to mankind in one way or other, is in serious risk of extinction over the next 2-3 decades. On realisation that the erosion of biodiversity may threaten the very extinction of life has awakened man to take steps to conserve.

In the last seven years, plans for biodiversity conservation have been developed by the WR1 and the IUCN with support from World Bank and other institutions. Basically, the conservation plan has an holistic approach and encompasses whole spectrum of biota and activities ranging from ecosystems at the macro level *in situ* conservation to DNA libraries at the molecular level conservation.

Plan For Action

A conservation of the biodiversity, in the immediate risk will be to devise and enforce time bound programme for saving plant and animal species as well as habitats of biological resources. Action plan for conservation, therefore, must be directed to:

- Inventionsation of biological resources in different parts of the country including the island ecosystems;
- Conservation of biodiversity through a network of protected areas including National Parks, Sanctuaries, Biosphere Reserves, Marine Reserves, Gene Banks, Wetlands, Coral Reefs, etc.;
- Rehabilitation of rural poor/tribals displaced due to creation of protected areas.
- Conservation of micro-organisms which help in reclamation of wastelands and revival of biological potential of land;
- Protection and sustainable use of genetic resources/germplasm through appropriate laws and practices;
- Control of over-exploitation through CITES and other agencies;
- Protection of domesticated plant and animal species in order to conserve indigenous genetic diversity;
- Maintenance of corridors between different nature reserves for the possible migration of species in response to climate, or any other disturbing factor;
- Support for protecting traditional skills and knowledge for conservation;
- Multiplication and breeding of threatened species through modern techniques of tissue culture and biotechnology;
- Discouragement of monoculture introduction; and
- Restriction on introduction of exotic species without adequate investigations.

India is the second most populous country, and therefore, any plan attempting at conservation must consider socio-economic development as the mounting human pressure threatens the biotic resources of the country. Furthermore, ours is predominantly an agriculture country, and hence, policy makers should realize that conservation and sustainable utilisation of biodiversity is the centre to all developmental planning.

CONSERVATION METHODS

(a) *In situ* Conservation

In situ conservation refers to protection zones and areas of high biological diversity. These areas, described as natural ecosystems, will protect species with minimum human interference. The buffer zones or seminatural ecosystems can allow for some human disturbance as long as the impact of humanity is not greater than any other factor.

For preservation of the endangered species, the only measure suggested is the strict protection against poaching of both vegetation as well as animal resources. Since most of the threatened organisms occur as components of biotic communities in open sites, restoring them in such habitats through judicious protection measures is required. For *in situ* conservation, the biosphere reserve offers the best site of natural conservation of threatened flora. Today, India has 75 national parks and 421 wildlife sanctuaries covering an area of about 1.4 lakh km^2 constituting more than 4% of the total geographic area of the country, and one-fifth of the forest area. The protected area includes 23 tiger reserves as well as 14 biosphere reserves.

India's efforts at countering the rapid erosion of biodiversity have been significant as it provides the world's largest network of protected areas *in situ* conservation. Also, population of the great Indian bustard, the tiger, the elephant, the Hangul, the crocodile, etc., have gone up. This shows that with proper implementation of conservation measures the loss of biodiversity can be compensated.

The Wildlife Institute of India has comprehensively reviewed the existing protected area network and highlighted the need to identify new protected areas in different parts of the country, in order to ensure representation of maximum wildlife habitats. It has made proposals to increase the existing network coverage in India to 147 national parks with an area of 49,435 km^2, and 519 wildlife sanctuaries with an area of 116,879 km^2 raising the coverage upto 5.06% of the total land area. The proposed network should be accepted.

The conservation efforts towards plant species have not been given adequate attention particularly of those which are of potential economic and scientific value. The other important aspect is to incorporate rehabilitative strategies for rare, threatened and endangered plant and animal species. In order to derive maximum benefit from the *in situ* conservation, incentive to grow domesticated economically important biota and develop herbal drug industry should be provided to the people who can, in particular, use wastelands of the country. One of the foremost requirements of *in situ* conservation is to prepare a comprehensive National Biological Inventory, Immediate steps are to be taken in this direction.

Scientific studies in regard to *in situ* conservation should focus on the following lines where very little data are available:

- Applied research for conservation of living resources;
- Interlinkages between plant and animal species;
- Quantitative assessment of the conservation status of the species;
- Successional status of key species in different ecosystems;
- Multiplication and restoration of endangered, rare and endemic species using biotechnology;
- Ecological restoration of degraded micro and macro-habitats;
- Identification of critical index species and their sensitive parameters;

- Assessment of the impact of exotic species on the ecosystem;
- Determination of the impact on the ecosystem of various activities in the protected areas;
- The possible climate change and its impact on biodiversity;
- Hydrological changes including surface run-off and percolation in the protected areas;
- Primary production and cycling of nutrients in the soil;
- Studies on satellite mapping of all protected areas; and
- Development of methodologies for classification of micro-habitats.

The important point in *in situ* conservation is that the forest trees, wild plants, wild animals and micro-organisms all occur together in an ecosystem. Therefore, if an attempt is made to conserve and enrich the ecosystem, much can be achieved in a single step. This would be particularly advantageous in tropical forests where many species occur in low densities and have a high degree of endemism. Obviously there is an urgent need to coordinate efforts on the ground level. This would save not only time and effort but also the scarce fiscal resources and infrastructure.

Local communities need to be involved in the management of biological resources and to benefit from their sustenance. This can be achieved by integrating rural development with ecosystem conservation. If this point is missed or weakened, all measures of conservation of biological resources would not help much.

To identify ecosystems that have been left out and are in urgent need of conservation, it is necessary to match the 12 bio-geographical provinces (viz. Ladakh, Himalayan highlands, Malabar rain forest, Bengal rain forest, Indus-Ganga monsoon forest, Assam-Burma monsoon forest, Mahanadian,

Coromandel, Deccan thorn forest, Thar desert, Lakshadweep Islands, Andaman and Nicobar Islands) with the present day protected areas network. From such a study there will emerge the additional areas which are in need of conservation. The process of identification of the additional areas must be based, among other things, on the following :

- Vavilovian centres of diversity of crop plants in the Indian region particularly with regard to wild ancestors of the crop plant genetic resources, non-crop plant genetic resources, and forest tree genetic resources;
- Wild relatives of livestock;
- Fish genetic resources;
- Fresh water systems (rivers and lakes);
- Marine fish and other economic sea animals, Mangrove and coral systems; Island ecosystems;
- Threatened/endangered biota including materials used for teaching; and
- Unique and fragile ecosystems, including hot spots (NE Himalayas) and endemic areas.

So far, the approach has been restricted to fauna, especially large mammals and big cats in particular, because they are at the top of the food chain. Such a restricted view must change in favour of a holistic one in order to enable the country to save as much of the biological wealth as possible.

(b) *Ex situ* Conservation

India has done commendably well as far as *ex situ* conservation of crop genetic resources is concerned. It has also taken up such work on livestock, poultry and fish genetic resources. However, there is need to develop facilities for long and medium term conservation through:

- Establishment of Genetic Enhancement Centres for producing good quality of seeds;
- Enhancement in the existing zoos and botanical garden network;

- Seed-gene banks
- Tissue culture gene banks;
- Pollen and spores banks;
- Captive breeding in zoological gardens; and
- *In vivo* and *in vitro* preservation

However, both *ex situ* and *in situ* conservation of forest trees and micro-organisms (except nitrogen-fixing blue-green algae) have not received much attention.

Ex situ and *in situ* conservation should be given equal importance as measures in biodiversity conservation. Release of genetically modified organisms should be regulated at national and international level, and there should be adequate dissemination of information about such release by the respective countries.

Pic. 19.1. Weather Vagaries: Diseases might be caused in the climate.

4

IMPACT OF BIODIVERSITY CONVENTION ON CONSERVATION

"The Convention Biological Diversity" was one of the foremost issues discussed at the Earth summit held at Rio de Janeiro (Brazil) between June 3 and 14, 1992. A ceremony to mark the opening of the convention on biological diversity took place in the afternoon of June 5. Feranando Collor, the President of the Federal Republic of Brazil was the first to sign the convention, followed by India, and 155 other nations. While major industrial powers including United Kingdom, Germany and Japan had decided to sign the convention, the United States refused to do so. The US has thus refused to 'own' the responsibility for causing environmental degradation and had escaped the obligation of providing additional funds and transfer of new technology. The US fears for opposing the convention on biodiversity are not justified, and its refusal to sign the document shows that its decision has been influenced by big industries in that country rather than by environmental concerns. One of the point argued by the US team at the negotiation table was that the patents right on new technology were owned by private industries and that fell within the domain of intellectual property right. But the developing countries led by India had put the counter argument that they did not have the necessary funds to invest in new technology at an exorbitant rate and under the convention, the US must provide necessary 'incremental costs' by way of additional contributions to the global environmental facility so that new and additional financial resources could be provided for that purpose.

The basis of the convention on biodiversity is the recognition of the fact that the animal and plant life on the earth is endangered by the over-exploitation and excessive emission of toxic gases by the industries in the developed world. The convention seeks to evolve a principle of 'global participation' and makes it obligatory on the participating countries to take corrective steps for the conservation of plant and animal life. This global agreement comes against a backdrop of dreadful destruction. The agreed text of the convention is appended at page.

The main features of the convention are:

- The authority to determine access to genetic resources rests with national governments and is subject to national legislation;
- The commercial benefits arising out of the use of biological resources of a country will be shared with that country on an equitable and fair basis;
- The access to the transfer of technology to developing countries will be provided under fair and most favourable terms mutually arrived at. In case of technology subject to patents such access and transfer shall be provided on terms which recognise and are consistent with the adequate protection of intellectual property rights.
- The developed countries will provide 'new and additional' financial resources to enable developing countries to meet the agreed full incremental costs to them of implementing measures which fulfil the obligation of the convention; and
- The extent to which developing country parties will effectively implement their commitments under the

> convention will depend on the effective implementation by developed country parties of their commitments under the convention related to financial resources and transfer of technology. The level of implementation by developing countries will also take into consideration the fact that economic development and poverty eradication are their primary and overriding priorities.

It may be mentioned that India has made it clear that biodiversity and sharing of biotechnology were inseparable issues. India has told the developed countries that if access to biotechnology or its products is going to be denied to the developing countries on grounds of protection of intellectual property rights, then such countries providing the genetic material would have little enthusiasm for conserving biodiversity.

Conservation In Context Of Biodiversity Convention

At Global Level

Apart from the various clauses committing countries to national and international conservation measures, the convention at global level rather appears to be political. This is apparent from the sabotaging act of the US who has refused to sign the biodiversity document. The objective of refusal was to globalise the natural resources of the developing countries while allowing its own resources to be ruthlessly exploited in the plea that these were privately owned. However, on this count, the developing countries are in the bargaining position as they have 90% of the world's total biodiversity resources. But to do this will require an unprecedented show of solidarity amongst the gene-rich tropical nations of the world, and cooperation between groupings ranging from regional bodies like SAARC to trans-continental for like G77.

Some of the important clause, which provide effective conservation measures at global level are discussed as follows:

1. Under the convention (Article 3) the national sovereignty has been accepted over biodiversity. This means

that access to plants, animals and micro-organisms are now subject to "mutual agreements' between the countries. For the first time, a provision has been made for checking one-way transfer of resources from developing countries to the developed countries. Mutual agreement' has yet to be defined. For instance, what would be the demand by the donor country in return for access to genetic resources:

- one time payment or continuing royalties from the profits gained by the receiving country,
- access to relevant technologies, including biotechnology or
- access to genetic products, for example hybrid seeds.

Whatever be the exchange, it should be in benefits of both donor as well as recipient countries.

2. The relevant and acceptable technologies and fund which the developing countries can get from the developed countries have to be explored. This is mentioned in the various provisions in the convention, including Article 16 which commits nations to "provide and/or facilitate access for and transfer to other contracting parties of technologies that are relevant to the conservation and sustainable use of biological diversity or make use of genetic resources and do not cause significant damage to the environment". Article 20 further provides for the transfer of "new and additional "financial resources from the developed countries to the developing countries, and for the setting up of a separate, democratically managed fund for biodiversity conservation."

It is not necessary that 'Western technology' is always promising. There are a number of ecologically sustainable skills and techniques in developing countries that need to be encouraged and revived, rather than being replaced by modern ones. This is not to say that no western technology is useful, but that its wholesale adoption could be counterproductive. Similarly, the flow of funds from the North will not necessarily enhance biodiversity conservation. A global biodiversity fund can be helpful only when it is democratically managed (unlike the donor-dominated, WB controlled international fund, the GEF), open to public scrutiny, and sensitive to the environmental context of the recipient country. Having got this accepted as the convention text, the Southern countries must continue pressing for such a fund.

Even after relevant technologies and biologically modified products are identified, the Southern access to them may be restricted by patents and other forms of IPRs. The Convention is not clear on this score. Article 16(2), for instance, subjects to technology transfer to existing IPRs. However, Article 16(5) states that countries should ensure that IRPs lire supportive of and do not run counter to the objectives of this Convention.

The developing countries can expose the IPR regimes being imposed by the developed countries as countering the biodiversity conservation interests. This is likely to be an uphill task, given the moves under GATT and related negotiations. Apart from mass movements, the only international tool we have to come iniquitous and destructive trends like GATT is the Biodiversity convention. Article 22(1), which appears to disallow "exercise of rights and obligations" under other international agreements could be used to this end.

Safeguards have to built against the misuse of biotechnology. Evidence shows that the potential of modern biotechnology is tempered by its capacity to induce biodiversity erosion, environmental and health damage, and greater economic and social inequality within and among the nations. The Convention is unsound in this respect, but Article 19(3) suggests that the countries consider a protocol on "safe transfer, handling and use of any living modified organism resulting from biotechnology that may have adverse effect on the conservation and sustainable use of biological diversity."

In the final analysis, the Biodiversity Convention is effective only if it is creatively and collectively used. Will all the communities from developing as well as developed countries now respond to the challenge to forge a civilisation in which humanity is at a peace with its 1.5 million fellow species.

At Country Level

India is fortunately placed in a position of advantage. Ours is a tropical country with a tremendous heterogenecity of environments ranging from tropical rain forests of Andaman and Arunachal Pradesh to the deserts of Rajasthan and Ladakh. It lies at the junction of the three biogeographical provinces of Africa, temperate Eurasia and the Orient.

As a result, it has rich biological heritage that qualifies it as one of the 12 megadiversity nations of the world.

The industrial nations, on the other hand, lie in temperate regions of the world, that are quite poorly endowed with natural diversity. Also, many of these countries have suffered from severe onslaught on Nature till mid 19th or early 20th centuries. As a result, while these nations are far ahead of the tropical world in technologies, the bulk of biodiversity resides in third world countries.

India is, in a way, a connecting link. We are not so rich in biodiversity as Colombia or Indonesia, nor so advanced technologically as Germany or Japan. But we possess both substantial levels of biodiversity and technological capabilities. So we must take the lead in steering the Biodiversity Convention in the direction of brighter scenario. Not only this, but also being signatory to the convention, India has moral binding to adopt conservation measures as provided in various clauses of the document. A few suggested measures are discussed as follows:

1. India has to review its development and agriculture policy to incorporate the conservation measures as suggested in draft of Biodiversity Convention. Article 6 of the convention states that each contracting party shall, in accordance with its particular conditions and capabilities integrate as far as possible and as appropriate, the conservation and sustainable use of biological diversity into relevant sectoral or cross-sectoral plans, programmes and policies while Article 14 requires that

each contracting and "introduce appropriate producers requiring EIA of its proposed projects that are likely to have significant adverse effects of biological diversity with a view to avoiding or minimising such effects. "

The agricultural policy should also be revised to incorporate conservation and upgradating of indigenous crops and livestock varieties perhaps by redirecting subsidies and incentives towards them. But it will require an uphill, long drawn struggle to get the minds of decision makers changed. Indeed, the recent economic policies, particularly the IMF-World Bank inspired liberalisations and structural adjustment process, may make this task virtually impossible. More than ever before, the clamour for foreign exchange is likely to lead to pressure to sell off our biological resources. An indication of this is the recent move by the Commerce Minister to reintroduce the export of plants and other products form the wild. The main hope is that the increasing number of mass movements and active participation of environmentalists' will pressurise the Government to change such destructive developmental policies.

2. An immediate step is to incorporate biodiversity concerns in the existing EIA procedures. A long pending proposal for making EIAs mandatory for both private and public sector development projects must be immediately notified.

India has to take fresh look at the conservation programmes, policies and laws. This follows the commitment in Article 6 of the convention-" that each contracting party shall in accordance with its particular condition and capabilities develop national strategies, plans or programmes for the conservation and sustainable use of biological diversity or adapt for this purpose existing strategies, plans or programmes" The 1992 National Conservation Strategy and Policy Statement on Environment and Development has to be followed up with a detailed biodiversity conservation strategy. One step towards this is the comprehensive status report on Indian Biodiversity, currently being prepared at the IIPA.

3. Another measure is to strengthen the laws on the biodiversity conservation. An assessment of over 40 Acts by the IIPA found serious gaps in conservation of domesticated biodiversity, restriction on introduction of exotica, and appropriate sharing of benefits of biodiversity use. The Government is considering a separate, over reaching legislation on biodiversity. However, there are certain difficulties in formulation of such legislation:

- How does one deal with both wild and domesticated biodiversity in a single Act?
- How does one legislate meaningful measures in the absence of even basic knowledge on biodiversity among the personnel concerned, including the judiciary?
- How do local community rights become the centrepoint of such legislation?

The conservation in India, as in the rest of the world, has focussed mainly on some glamorous animals (as wild cat, elephants, etc.) while ignoring the rest of the biological diversity. Particularly serious is the neglect of insect and other invertebrate life, and a majority of plant species.

Regarding the involvement of local communities in the conservation programme, Article 80 states that "each contracting party shall as far as possible and as appropriate subject to its national legislation, respect, preserve and maintain knowledge, innovations and practices of indigenous and local communities embodying traditional lifestyles relevant for the conservation and sustainable use of biological diversity and promote their wider application with the approval and involvement of the holders of such knowledge, innovations and practices and encourage the equitable sharing of the benefit arising from the utilization of such knowledge innovation and practices ". "Unfortunately, the villagers living in or around areas of rich diversity have been adversely affected by the curtailment of traditional rights to land and resources. This is not only violation of human rights but also a suicidal approach to conservation. No protected area is going to survive for long in the midst of a hostile and hungry population. The only way to overcome this hurdle is to ensure involvement of local communities in conservation programme, by guaranteeing their land and forest rights, and providing them a stake in the sustainability of their natural surroundings.

5

GLOBAL GENE BANK

The green revolution and genetic selection in plant and animal breeding depends upon diversity among genetic types. Development in biotechnology, notably in recombinant DNA, is limited, by the availability of genetic material. Further, the emergence of biotechnology as a powerful tool of the change in the world makes the preservation of genetic diversity indispensable for conducting hybridization programmes. Gene banking provides a method of conservation of genetic diversity, in addition to *in situ* and *ex-situ* conservation methods. In fact, it is not a new concept, but simply long-term storage of seeds at sub-freezing temperatures as gene banks, which are the means to ensure secure access to scarce varieties. The National Bureau of Plant Genetic Resources, New Delhi has already developed cold gene storage modules, where seeds can be sent for maintaining genetic diversity. To preserve the diverse genetic resources, particularly of threatened species and those the seeds of which are not viable for longer periods, banking is one of the most suitable methods for conservation. This calls for an international consensus on establishing global network of gene banks.

Soil is a natural gene bank to which seeds are added year after year. When a wild plant produces seeds, not all the seeds will grow in the following season. Much of the seeds drop to the ground and wait there patiently for right conditions to permit germination.

In fact, man-made gene bank is a kind of freezer that preserves seeds, pollen and spores. Chest freezers are the best choice for small gene banks. Upright freezers, on the other hand are not suitable as the cold air rushes out when the door is opened. Chest freezers with a temperature of-15°C are adequate for this purpose. Operating and maintenance of the equipment does not require extensive knowledge of specialised training, nor does it require heavy investment. This simplicity of a gene bank makes it viable project for plant enthusiasts, ecologists and NGOs who are interested in biodiversity conservation.

Theory Of Gene Banking Preservation

The theory behind gene banks is really quite simple, i.e., seeds of many plant species remain viable longer when their moisture is reduced and stored at low temperature. Cryogenic preservation is no more than suspended animation at sub-freezing temperatures and low moisture content. The process requires little more than a household chest freezer, storage containers made of glass, metal or plastic, a simple desiccant or drying agent to dehydrate the seeds.

Seeds are tiny organisms that stay alive so long as there is a food reserve inside the seed to charge vital chemical reactions that keep the seeds alive. These chemical reaction rates are dependent on the ambient temperature and the water content of the seed. The higher the temperature, the faster the reactions will take place and in the less time the food reserves will last. The converse is also true, i.e. for every 5°C drop in temperature, the life of the seed will double. For example, onion seeds with 10 per cent moisture are viable for 16 weeks at 35°C but will live for 78 years at 0°C. Dropping the temperature down to –15°C would increase longevity to 624 years, provided the ice crystals formed at 10 per cent moisture level do not harm the seeds when frozen. Clearly, if seeds could be stored at sub-freezing temperatures, they might be viable for centuries. Cryogenic theory, thus, suggests that extremely low temperatures allow seeds to remain viable for many centuries, but such tremendous life-spans practically may not necessary be achieved.

Too much water in a seed will form ice crystals as the temperature drops below the freezing level. Seed with a high water content is plagued by large ice crystals that rupture individual cell membranes and destroy the seed. Many seeds, however, have a reduced water content and so can be frozen soon after they have ripened without danger of rupture. On the other hand, some seeds require extra care to reduce their water content. While fleshy or oily seeds cannot easily be frozen and must be preserved by some other methods.

Normally, the life-span of a seed doubles with each 1 per cent decrease in water content. A minimum of 4 per cent water content appears to be required by seeds to stay alive. International standards for long-term seed storage suggest that an average seed should contain five per cent water and be stored at-18°C. When both temperature and water content are reduced, the two factors multiply. For instance, when temperature drops 5°C and water decreases by 1 per cent, the seed lives four times as long.

Institution Of A Gene Bank

It is accomplished in the following steps:

1. Collection and drying of seed sample

Before we go for collection, the first step is to ascertain the viability of seeds as well as the size of seed sample. A variety of stains and tests can be used. The simplest method is used to determine the percentage of seed germination. Botanists usually strive for about 95% germination. Dye 2, 3, 5-triphenyltetrazolium chloride is the best stain for viability test of seeds. Viable seeds respire and thus are capable of changing colourless tetrazolium dye into highly coloured (red) compound by chemical reaction.

The species which have low rate of germination are the endangered species. The size of the seed sample depends upon its use. When sample is required to conserve the whole germplasm, that is full range of variation within the population, then the sample should contain roughly 10,000 seeds. On the contrary, a small sample of 20 to 100 seeds might be sufficient, if it is required to retrieve a few plants for illustrative specimens, breeding experiments or genetic research. The Gene banker is advised to have the seed samples in duplicate or even in triplicate to facilitate the easy removal of desired amount of seeds from the freezer whenever is needed.

Store clean seeds (without chaff and stalks) only and not the capsules to save the limited space available in the freezer and make it easy to inspect and observe. For collection of seeds, harvest the pods as soon as they start to ripen or split. Put the seeds in a container, and expose them to air for few weeks. The seeds held in the gene bank at the Royal Botanic garden Kew, are dried in a conditioning room with a temperature of 16°C and a relative humidity of 14 per cent.

Seeds can be dried simply by storing them for several days in a closed container with a desiccant/drying agent. The most familiar drying agent is Silica Gel. Another product which readily absorbs moisture from the air is Drierite, a calcium salt. Drierite comes with a colour indicator; when it turns blue, it can still absorb moisture; when it turns pink, it is hydrated and can absorb no moisture. It is important to test the drying agent before using it to dry seed. A handy test for assessing its freshness is to use a strip of filter paper saturated in a solution of cobalt chloride. The paper strip is blue when dry, and pink as it becomes moist. When you find that the drying agent is stale, rejuvenate it by baking it on a Teflon-coated baking sheet for two or three hours at 60°C or higher.

When the drying agent is ready to use, find any container that can be sealed. About 2.5 cm of drying agent should be placed at the bottom of the container and covered with either cheesecloth or wire gauge, permitting air to reach the drying agent. Open containers holding seeds should be put into the desiccator and the desiccator is sealed. Open it at least once a day to stir the seeds. During the first two days seeds will lose most of their water content; a little more is lost during the next five or six days. This is the most practical method of drying seeds for gene bank.

Higher temperatures-35°C or above-accelerate the metabolic reactions and so affect the vitality of seeds. It is, therefore, desirable to dry the seeds at gentle room temperature.

Freeze-drying under vaccum has been considered the most suitable technique for drying seed sample. However, it requires extra care as a result of formation of ice crystals at freezing temperature and strong vaccum used to dry the material.

Storing

Containers made up of metal, glass, and plastic can be used for storing seeds. Large samples of big seeds, such as beans and peas, store very well in metal containers. Glass containers, the second choice, are more fragile but this method is relatively cheap. Inexpensive disposable glass test tubes are easy to get from scientific or medical supply houses. End of the tube can be sealed with a propane torch, without significantly heating up the seed sample. With glass container it is easy to view the frozen sample. The third type of containers are envelopes made of foil-laminated plastic that can be heat-sealed. These envelopes have an inner lining of polyethylene and an outer lining of aluminium or some other metal. While sealing, vaccum device can be applied.

Labeling

It is important to label all samples. Care should be taken to use labels that must resist sub-freezing temperatures, for long periods of time. It is adviced to use ink that will not fade soon. Further dual system of labels is preferred. One label should be placed on the outside of the sample and another should be placed inside. As an extra precaution, a record of the sample's self position should be maintained in a book or minicomputer. The name of the sample should be imprinted onto a metal foil label. There is also an idea to number the samples and to have the permanent record of identification.

Records should be precise and document, the quality and viability of the seeds stored.

Pollen and Spore Banks

Preservation of pollen and spores is of significant value for conservation of biodiversity of important flowering and spore bearing plants. The procedure for institution of pollen and spore banks is almost similar to that of gene banks.

Cryogenic technique is useful in preserving pollen from flowering or cone bearing plants, and spores from non-flowering plants, such as ferns and mosses. Pollen are so tiny that thousands and even millions can be stored in a very small vial. Pollen preservation is thus advantageous over seeds, preservation as it gives opportunity to preserve the full range of variation within the population in a very simple manner. However, plant scientists are able to make use of pollen form certain plant species only to grow them into whole plants. We expect that development in science and technology, in near future, will solve this genuine problem. Pollen can also be used in hybridization programme whenever needed. When seeds are used to develop hybrids they must be retrieved from the bank and then planted. It takes several years for the plant to be useful in hybridization. On the other hand, pollen can be used 'as is', it is much more economical and efficient. A pollen bank can be an extremely powerful tool in plant breeding since it frees breeders from the tyranny of time. Also, it is useful in self-sterilised plant species. Pollen should be collected from fresh flowers only.

Fern spores, which are similar in size to pollen, can be treated just like pollen. Even the spores of tropical ferns like staghoms, platycerium species, can be germinated after drying and freezing. Pollens are stored in gelatin capsules, the type used to hold medicine. The stamen, which contains pollen, is removed from the flower, inserted into the capsule and shaken several times to deposit the dust-like pollen on the capsule wall. The stamen is then withdrawn. The name of species and date of collection may be written directly on the capsule with a waterproof laundry marker. Dry the pollen by exposing the capsules to the air inside a frost-free refrigerator for 24 hours. Store the capsules so exposed inside individual plastic containers that have a little Dnente in them to prevent from moisture moving through the gelatin and then place these containers in the bank. Relevant data are written on the wall of the plastic container.

Gene Banks At Global Level

The National Seed Storage Laboratory

The United States' National Seed Storage Laboratory (NSSL) is located at the University of Colorado in Fort Collins. It does have some areas for sub-freezing storage, but most of the accessions are stored at 5°C.

The Vavilov Institute

N.I. Vavilov Institute in Leningrad is the primary Russian gene bank. It has 25 experimental stations around the country. The facility houses have an enormous collection of seeds and living plants, ranging from wild species and primary crops to breeding lines of modern crops, though most of its work is in plant introductions. In 1975, the institute's collection had more than 200,000 accessions due to its aggressive collection programme. A large seed exchange programme with foreign countries has been established and a large seed bank at Kubran has been developed.

The Izmir Centre

The Izmir Centre, one of the most important centres of biodiversity, was established in Turkey in the mid-1960s. The seeds of the area were sampled extensively and the center proved to be very important for cereals-particularly wheat- and for forage legumes such as alfalfa. Within ten years, more than 13,000 seed collections from Turkey itself and at least 25,000 collections from other countries had been assembled. The collections at the institute are vital. Political instability in the region may affect the long-term usefulness of the centre.

The International Rice Research Institute

The International Rice Research Institute (IRRI) at Los Banos in the Philippines was started in 1960 to collect rice varieties for breeding. It is one of the most successful examples of a gene bank. It has about 60,000 rice accesssions. Some of the nee species have been developed into high-yielding dwarf nee varieties that now feed much of the world. Today scientists at IRRI are attempting to breed varieties resistant to insects and disease, as well as have the ability to grow in drier climates or brackish water. Despite the huge number of accessions already located at the IRRI, some experts fear that valuable wild nee strains that would have contributed much to the new breeding stock already have been lost.

An important problem for the IRRI is the regeneration of seeds from time to time, particularly if they are not being stored at sub-freezing temperatures due to not working of refrigeration equipment in the humid tropics. Regeneration should occur at about 25 to 30 years intervals to ensure viability. Another problem is that Los Banos is susceptiable to earthquakes and typhoons. However, the region is ideal for year-round rice growing.

The Royal Botanic Gardens at Kew

The Royal Botanic Gardens at Kew is the first serious gene bank for wild species, and is the model for other gardens, Seed units were set up at the Jodrell Laboratory at new and at the satellite garden at Wakehurst Place, Sussex. Seeds for the bank were obtained from cultivation at Kew as well as collection from other botanic gardens.

The Bank in Irvine

The seed bank at the University of California Irvine (UCI) was primarily aimed at to collect South African bulbous and cormous plants. Families Iridaceae (the iris family) and Liliaceae (the lilies) were selected for this particular work. Workers at UCI, rather than collecting species at random, focussed their attention on these two groups.

The gene bank at UCI has several components: a living, growing plant collection, and cryogenic seed and pollen banks. The gene bank was set up not only to preserve endangered species but also to demonstrate that a small garden can set up a cryogenic gene bank which is economical and does not require technical expertise.

Other wild banks

Gene banks to preserve endangered and regional plant species have been initiated by the University of Dubin, the Polish Academy of Sciences and the Polytech University in Madrid. This was a cooperative effort among countries in the mediterranean region aimed at saving wild species. Botanists working on Mediterranean flora have organised a group called Organization for the Phyto-Taxonomic Investigation of the Mediterranean Area (OPTIMA). The important functions of OPTIMA are to list threatened plant species and to establish a network of seed banks.

Saving Clones

Many fruit trees and bushes are selected clones that would not come true from seed. The only method to preserve them is with collection of mature trees.

6

IMPORTANCE OF BIODIVERSITY

Impact of Biodiversity

The well being and survival of human beings are dependent on millions of species of plants, and microbes. These include species and varieties of crops, livestock and wildlife. Biological diversity is part of our daily lives and livelihood and constitutes resources upon which families, communities, nations and future generations depend. Biodiversity has numerous values in agriculture, medicine, food and industry. Biodiversity maintains ecological balance and evolutionary processes and has spiritual, cultural, aesthetic and recreational values. The services rendered by the ecosystem include pollination, cycling of biogenic salts and elements, soil maintenance, decomposition, self purification and climate regulation.

Millions of species which constitute biodiversity have their own intrinsic value as recognised by religions and faiths in India. Traditionally, Indians have celebrated this diversity of life. Conservation and sustainable use of biodiversity is ingrained in the Indian ethics and culture. Even today, millions of people derive their daily sustenance from forests, rivers, grasslands and seas. The practices and beliefs of forest dwelling tribes, nomadic communities, fisherfolk and farmers remain closely interwined with the biological diversity they live amidst.

Excessive use coupled with lack of awareness about the actual and potential value of biological resources has led to the loss of biological diversity. It is evidenced in the extinction of some wild species like Cheetah and some domesticated plant varieties and animal breeds. Rapid depletion in biodiversity is largely due to disempowerment of local communities, commercial exploitation, fragmentation and degradation of habitats. This loss of biodiversity is being felt through greater water shortage, lack of fodder, fuel, and other livelihood resources. And with this, there has been serious erosion in the knowledge base of people, especially the men and women of local communities and of cultural values linking people with nature.

General Introduction

The term biodiversity is formed by a contraction of the term biological diversity. The term biodiversity was coined by Walter and Rosen in 1985. There are many definitions available for biodiversity. Biological diversity means the variability among living organisms from all sources including inter alia, terrestrial, marine and other aquatic ecosystems and the ecological complexes of which they are part, this includes diversity within species, between species and within ecosystems.

Biodiversity refers to the variety among living organisms and the ecosystem complexes in which they live. It includes forms of diversity right from the molecular unit to the individual organism and biospheric levels. In the simplest sense, biodiversity may be defined as the sum total of species richness, i.e., the number of species of plants, animals and micro-organisms inhabiting in a given habitat.

Types of Biodiversity

Broadly speaking, the term biodiversity includes three different but closely related aspects. They are genetic, species and ecosystem level diversities.

Genetic Diversity

It refers to the variation of genes within species. This constitutes distinct population of the same species or genetic variation within population or varieties within a species. Genetic diversity resides in the variation in the sequence of the four base pairs, such as components of nucieic acids constitute the genetic code. Other kinds of genetic diversity can be identified at all levels of organizations, including the amount of DNA per cell, chromosome structure and number. New genetic variations arise in individuals by gene and chromosomal mutations, and in organisms with sexual reproductions, these variations are spread through the population by recombination.

Only a small fraction of the genetic material (less than 1%) of the higher organisms is outwardly expressed in form, function and variation. The purpose of junk DNA which not expressed is still unclear. It is estimated that there are 10 different genes distributed across the world's biota. These genes do not make an identical contribution to overall genetic diversity. These genes which are responsible for fundamental biochemical reactions, undergo very little change through evolution in different taxa. For example, in the nuclear protein (histories) the sequences are conserved all through the different stages of evolution both in plants and animals.

Species Diversity

It refers to the variety of species within a region. Such diversity could be measured on the basis of number of species in a region. Perhaps, it is customary to refer the living world in terms of species, hence, the term biodiversity is commonly used as a synonym of species diversity. It refers to species richness, in terms of number of species in a site or habitat. Discussion of global diversity is typically represented in terms of global number of species in different taxonomic groups. An estimated 1.5 to 1.7 million species have been identified and described to date. The total existing organisms on the earth at present vary from five to thirty million. A conservative working estimate suggests there might be 12.5 million species.

Ecosystem Diversity

In a ecosystem, there may exist different landforms, each of which supports* different and specific vegetation. Ecosystem diversity in contrast to genetic and species diversity is difficult to measure since the boundaries of the communities which constitute the various subecosystems are elusive. Ecosystem diversity could best be understood if one studies the communities in various ecological niches within the given ecosystem, each community is associated with definite complexes. These complexes are related to composition and structure of the biodiversity.

Global Biodiversity

We believed that there may be 5 to 30 million species of organisms exist on our earth. These include 3,00,000 species of green plants. 8,00,000 species of fungi, 40,00,000 species of insects, 3,60,000 species of micro-organisms and many invertebrates and vertebrates (Table 22.0). According to some recent estimates, the number of insects alone may be as high as 10 million, but many believe that it is more likely to be around 5 million. Global diversity reserves and country-wise distribution of animal species are shown in Tables 22.1 and 22.2 respectively.

Table 22.0. Global Distribution of identified species

Sl. No.	*Types of Organisms*	*Number*	*Percentage*
1.	Plants	3,00,000	20
2.	Vertebrates	40,000	3
3.	Insects	8,00,000	53
4.	Microbes	3,60,000	24

The tropical forests are regarded as the richest in biodiversity. Scientists are of the opinion that whatever be the absolute numbers, more than half of the species on the earth live in moist tropical forests which is only 7% of the total land surface. Insects (80%) and fungi make up most of the species. For instance, from a single tropical leguminous tree 43 ant species belonging to 26 genera have been retrieved.

This approximately equals the ant diversity of all the British Isles.

The following reasons have been put forward with regard to the high species diversity in tropics:

- In tropics, conditions for evolution are optimum and for extinction are meager.
- In tropics, species diversity was conserved over geological time. This is because of low rates of extinction prevailing there.
- Biological diversity is the result of interaction between climate, organisms, topography, parent soil materials and the heredity. The tropical region is the ideal place for such an interaction.

Table 22.1. Detailed estimation of number of species spread worldwide

Sl. No.	*Group*	*Number*
1.	Bacteria	3600
2.	Blue greens	1700
3.	Fungi	46983
4.	Bryophytes	17000
5.	Gymnosperms	750
6.	Angiosperms	250000
7.	Insects	750000
8.	Sponges	5000
9.	Crustaceans	9000
10.	Molluscs	38000
11.	Starfishes	50000
12.	Fishes	6100
13.	Amphibians	19056
14.	Reptiles	6300
15.	Birds	9035
16.	Mammals	4008

Biodiversity in India

The Indian subcontinent (8°-30' N and 60'-97.5' E) having a geographical area of 329 million hectares is quite rich in biodiversity with a sizable percentage of endemic flora and fauna. This richness in biodiversity is due to immense variety of climate and altitudinal conditions coupled with varied ecological habitats. These vary from the humid tropical Western Ghats to the hot desert of Rajasthan, from the cold climate of Laddakh and the Leh mountain of Himalaya to the warm coasts of peninsular India.

Table 22.2. Countrywise distribution of total number of animal species (upto 10th rank)

Country	*Mammals*	*Country*	*Birds*
Indonesia	515	Columbia	1921
Mexico	449	Peru	1703
Brazil	428	Indonesia	1519
Zaira	409	Ecuador	1447
China	394	Venezuela	1275
India	372	Bolivia	1250
Peru	361	Malaysia	1200
Columbia	359	India	1200
Uganda	311	China	1195
Tanzania	310	Brazil	672
Country	*Reptiles*	*Country*	*Amphibians*
Mexico	719	Brazil	516
Australia	686	Columbia	407
Indonesia	600	Ecuador	358
Brazil	467	Mexico	282
Columbia	383	Indonesia	280
India	353	China	265
Ecuador	345	Peru	251

Peru	297	Zaira	216
Malaysia	294	USA	216
Thailand	242	Venezuela	197

Country	*Butterflies*	*Country*	*Angiosperms*
Indonesia	121	Brazil	55000
China	104	Columbia	45000
India	77	China	27000
Brazil	74	Mexico	25000
Myanmar	68	Australia	23000
Ecuador	64	S. Africa	21000
Columbia	59	Indonesia	20000
Peru	59	Peru	20000
Malaysia	56	India	15000
Mexico	52	All others	2000

The country has over 1,15,000 species of plants and animals already identified and described. In addition, the country is very important centre of biodiversity and the place of origin of over 167 important cultivated plant species, and some domesticated animals. To name a few, the following crops arose in the country and spread throughout the world, namely rice, sugarcane, Asiatic vignas, jute, mango, citrus, banana, several species of millets, cucurbits, some ornamental orchids, several medicinals and aromatic plants. In fact, our country has been recognised as one of the world's top 12 megadiversity nations. This region is also a secondary centre of diversity for grain amaranthus, maize, red pepper, soyabean, potatoes and rubber plant.

There are several examples of plant germplasm from India contributed to plant improvement. Two cases are well known:

1. Nearly 20 varieties of rice contain useful genes from wild rice of Kerala, which are responsible for resistance in cultivated rice.

2. The contribution to the cantaloup (musk melon) industry (14, (XX) ha) of California by powdery mildew-resistant genes is from the wild musk melon of our country.

Among the flora, the country can boast of 45, 000 species which accounts for 15 per cent of the known world plants. The distribution of the recorded flora of the country is shown in Table 22.3. Of the 15,000 species of flowering plants, 35 per cent are endemic and located in 26 endemic centers. Among the monocotyledons, out of 588 genera occurring in the country, 22 are strictly endemic. The family Poaceae has the highest endemism both by genera and species. The north eastern region could boast of being unique treasure house of orchids in the country, the abode of about 675 species out of 1,000 available in the Indian peninsula and against 17, 000 species the world over. The important Indian orchids include Paphiopedilwn faireyannm (Lind.) pfitz., Cymbidium aloiflinni sw. and Aeridcs crispnin Lind., etc.

Our country is also rich in faunal wealth. The country has nearly 75,000 animal species, about 80 per cent are insects. The distribution of major animal groups are shown in Table 22.4. In animals, the rate of endemism in reptiles is 33 % and in amphibians is 62 per cent. Further, there is wide diversity in domestic animals, such as buffaloes, goats, sheep, pigs, poultry, horses, camels and yaks. Domesticated animals too have come from the same cradles of civilization as the major crops.

Table 22.3. Distribution of Plants described already

Group	*Number of Organisms*	*Percentage*
Flowering plants	15000	33
Pteridophytes	600	1
Fungi	20000	45
Bryophytes	2700	6
Lichens	1600	4
Algae	5000	11

Table 22.4. Major Animal Groups of India and the World

Group	*Number of Spp*	
	World	*India*
Mammals	4231	372
Birds	12450	1200
Reptiles	6300	435
Amphibians	4184	181
Fishes	23000	2000
Insects	800000	60000
Molluscs	100000	5000

There are no clear estimates about the marine biota, though the coastline is 7,000 km long with a shelf zone of 4,52,460 sq. km and extended economic/one of 20,13,410 sq. km. There is an abundance of seaweeds, fish, crustaceans, molluscs, corals, reptiles and mammals. Information regarding other flora and fauna are fragmentary. Hundreds of new species may be present in our country awaiting discovery. The Western Ghats in peninsular India, which extend in the southern states, are a treasure house of species diversity. Out of the described 15,000 species of the flowering plants in India, about 5,000 species occur in the Western Ghats of Kerala, 235 are exclusive to this region. It is estimated that almost one-third of the plant varieties found in India have taken refuge in Western Ghats of Kerala alone.

Species Diversity and Ecosystem Stability

Species diversity or diversity of genes within species increases its ability to adapt to adverse environmental conditions. When these varieties of populations of these species are destroyed, the genetic diversity within the species will diminish to a minimum. In many cases, habitat destruction has narrowed the genetic variability of species lowering the ability to adapt to changed environmental conditions. Finally, this diminished genetic variability may lead to ecosystem instability. In other words, the greater the variability of the species, the more is the ecosystem stability.

Ecosystem stability has been considered to be related to the cycling and recycling of nutrients, which in turn, increases the efficiency of the resource use in the ecosystem. The high species diversity thus, may be instrumental in cycling and recycling of nutrients and thereby achieving the stability of the ecosystem. The survival and well-being of the present day human population depends on several substances obtained from plants and animals. The nutritional needs of mankind are also met by wild and domesticated animal and plant species. Indeed, the biodiversity in wild and domesticated form, is the source for most of humanity's food, medicine, clothing and housing, much of the cultural diversity and most of the intellectual and spiritual inspiration. It is believed that 1/4th of the known global diversity, which might be useful to man in one way or other, is at serious risk of extinction. This calls for an integrated approach for conserving global biodiversity.

Establishment of nature reserves and biospheres with lot of biophysical variability, maintenance of corridors with different nature reserve for the possible migration of the species in response to climate change, etc., are the immediate steps to be taken for conserving the biological diversity. An international consensus on establishing global network for gene banks, microbiological resource centers, and marine parks is also important. At the same time, conservation must be coupled with socio-economic development, especially in countries where population pressure threatens the national biotic resources.

Measures of Biodiversity

Alpha Biodiversity (within the habitat)

No matter now the data is finally represented, biodiversity eventually boils down to the censusing of individuals. An ornithologist typically walks along a transact and records the number of different birds found in that area. This method of recording is essential in biodiversity assessment. But having recorded the number of species and their abundance, the ornithologist has to use certain methods to be able to compare across different areas. The most commonly used techniques are Diversity indices. There are different approaches to calculate these indices, each with its own assumption and biases, hence, care should be taken to determine which index is best suited for the purpose.

Site A and Site B both have the same number of species and the total abundances are also equal. However, Site B is more diverse because the species are more evenly distributed. The diversity index incorporates both the number of species (i.e. Species richness) as well as their abundance. Such an index is called a Heterogeneity Index. The most commonly used index, the Shannon-Wiener index, is derived from information theory. Some of the diversity indices are explained in this chapter.

1. *Shannon-Wiener Index*

(a) Formula

$H^1 = -\Sigma_{p/i} 1 np/i$

Where P_i = proportion of individuals of the species.

(b) Interpretation

This index assumes that individuals are randomly sampled from an 'Indefinitely large' population. The index also assumes that all species are represented in the sample. It is moderately sensitive to sample size and places more weightage on richness, i.e., if there are rare species recorded, the index tends to give a higher value. The value obtained from a sample is in itself of no significance. The index becomes useful only while comparing two or more sites.

2. *Simpson's Index*

(a) Formula

Simpson's Index = 1-D

Where $D = \Sigma pi^2$

and *pi* is the proportion of individuals of the ith species.

(b) Interpretation

Simpson's index is heavily weighted towards the most abundant species in the sample and is less sensitive to species richness. It is moderately sensitive to sample size.

3. *Fishers Alpha*

(a) Formula

$S = \propto (1 + N/\propto)$

Where S is the number of species in a sample.

N is the number of individuals.

$\propto$, the index of diversity, is computed by an iterative procedure.

(b) Interpretation

The procedure for fitting the model is to calculate the number of species expected in each abundance class and compare that with the number of species actually observed using a goodness of fitness (x: or G test). This index sensitivity to sample size is low and gives more importance to richness.

4. *Brillouin Index*

(a) Formula

HB=(In N! Σ 7 Y-In n!)/N

Where N is the total number of individuals and n_i is the number of individuals of ith species.

(b) Interpretation

This index describes a known population, there is no room for uncertainty while using this index. It places more emphasis on species richness and is moderately sensitive to sample size. However, calculating the Brillouin Index can be time consuming and complex and can give misleading answers due to its dependence on sample size.

5. Avalanche Index

This index was recently introduced in an attempt to capture the taxonomic and ecological diversity that is ignored by indices commonly used by ecologists worldwide.

(a) Formula

$$A_i = \sum_{i=1}^{n}\sum_{j=1}^{n}(p_i\, d_{ij}\, p_j)$$

Where P_i and P_j are frequencies of ith and j^{th} species and

$$d_{ij} = \left[\sum_{k=1}^{c} = \left(X_{ik} - X_{jk}\right)^2\right]^{1/2}$$

Where X_{ik} and X_{jk} are the values of the ith and j^{th} species for k^{th} character and c is the number of traits for which information is available.

(b) Interpretation

In addition to using species richness and abundances, this index also uses the biological and ecological differences among species comprising a community. This index requires that the exact taxonomic position of the species in question need to be known as its eventual value depends on this feature (Ganeshilah et al., 1997).

Beta Diversity (between habitat diversity)

Beta diversity refers to the comparison of diversity between habitats. Beta diversity can be defined as a measure how different (or similar) a range of habitats or samples are in terms of the variety (and sometimes the abundances) of species found in them (Magurran, 1988). One common approach to Beta diversity is to look at how species diversity changes along a gradient.

As in the case of Alpha diversity, there are many measures of Beta diversity. These are commonly termed as similarity indices. Some of these indices are:

1. Morislta-Horn Index of Similarity

(a) Formula

$C_{ni} = \{2\Sigma a_{ni} b_{ni}\}/\{(da + db)\ aN.bN\}$

Where

aN is the total number of individuals in Site A and a_{ni} is the number of individuals in the ith species in A and da= $(\Sigma\ an^2_i)/aN^2$.

(b) Interpretation

This index is not strongly influenced by either species richness or sample size. However, it is highly sensitive to the abundance of the most abundant species. The value obtained from the index ranges from 0 to 1: the value '0' implying that the two samples are totally dissimilar and T implying that they are totally similar.

2. Jaccard's Index of Similarity

(a) Formula

$Cj = J/(a+b-j)$

Where J is the number of species found in both sites and a is the number of species in Site A and b is the number of species found in Site B.

(b) Interpretation

This index assumes a value of T between sites of complete[1] similarity and '0' for sites of complete dissimilarity. This index is easy to calculate, but the disadvantage is that the coeffii units take no account of the abundances of species.

3. Sorenson's Quantitative Index of Similarity

(a) Formula

$C_N = 2_{JN}/(aN+bN)$

Where aN is the total number of individuals in Site A, bN is the total number of individuals in Site B, and JN is the sum of the lower of the two abundances recorded for species found in both sites.

(b) Interpretation

This version of Sorenson's index was modified to take into account the abundances of the species as well. This index also returns a value of '!' for sites of complete similarity and a value of '0' for sites of complete dissimilarity.

More information on the above indices and their measurements can be found in Magurran (1988).

Gamma Diversity

The sum of alpha and beta diversities of the ecosystems is an expression of the biodiversity of a landscape, which is considered as Gamma Diversity (Daniels et al., 1996)

The Species-Area Relationship and The Theory of Island Biogeography

There is a good correspondence between the number of species and the extent of area they are found in. Roughly speaking, larger the area, more the number of species found in that area MacArthur and Wilson (1967) took this idea a step forward in their *equilibrium theory of island biogeography.* The main point of this theory is that the number of species on an island is determined by a balance between immigration and extinction, and that this balance is dynamic, with species continuously going to become extinct and being replaced by the same or a different species (Townsend et al., 2000).

Some of the predictions made by the theory are:

1. The number of species on an island should eventually become roughly constant through time.
2. Large islands should support more species than small islands.

After the advent of the theory, a number of experiments have been conducted to determine the exact nature of the species-area relationship. While the original theory specified islands, similar work has been carried out in mainland areas where a certain habitat type is enmeshed in a matrix of the other habitat type. An example for a virtual island is the grassland-shola matrix found in South Indian forest ecosystems. (Caneshaiah et al., 1997).

Hotspots of Biological Diversity

The simplest method of suggesting target areas for conservation action is to identify countries with the highest numbers of species (greatest species richness) and those countries with highest species richness are called as megadiversity countries (Myers, 1988).

Though life teems in every corner of the world, this is not spread uniformly. Certain regions, such as the tropics harbour a multitude of species. Areas where high levels of species richness, threat and endemism coincide are termed as hot spots (Myers, 1988, 1990; Gaston and Williams, 1995; Myers et al., 2000). There are 25 recognised hotspots (Fig. 22.1 and Table 22.5) in the world (Myers et al., 2000). These areas contain about 133,149 plant species (44% of the world) and 9,645 vertebrate species in 2.1 million square kilometers or 1.4 per cent of the world's land area. Other taxa are also well represented in these areas (Tables 22.5 and 22.6). These hotspots harbor many rare and endangered species. Two criteria help in defining hotspots namely rich endemism and the degree of threat. To qualify as a hotspots an area must contain at least 0.5 per cent or 1500 of the world's 300 000 plant's species as endemics (Myers et al., 2000).

Loss of Biodiversity

The loss of biological diversity is a global crisis. There is hardly any region on the earth that is not facing ecological catastrophes. Of the 1.5 million species known to inhabit the earth (humans are just one of them), one-fourth to one-third is likely to become extinct within the next few decades. Biological extinction has been a natural phenomenon in geological history. But the rate of extinction was perhaps one species every 1000 years. But man's intervention has hastened the extinction rates. Between 1600 and 1950, the rate of extinction went up to one species every 10 years. Currently, it is perhaps one species every year.

Table 22.5. Showing different Hot Spots of the World

Sl. No.	*Hotspots*	*Plant species*	*Endemic Plants*	*% of global Plants*	*Verte-brate Species*	*Endemic verte-brates*	*% of global verte-brates*
1.	Tropical Andes	45000	20000	6.7	3389	1567	5.7
2.	Mesoamerica	24000	5000	1.7	2859	1159	4.2
3.	Caribbean	12000	7000	2.3	1518	779	2.9
4.	Brazil's Atlantic Forest	20000	8000	2.7	1361	567	2.1
5.	Choc/Darien/ Western Ecuador	9000	2250	0.8	1625	418	1.5
6.	Brazil's Cerrado	10000	4400	1.5	1268	117	0.4
7.	Central Chile	3429	1605	0.5	335	61	0.2
8.	California Floristic Province	4426	2125	0.7	584	71	0.3
9.	Madagascar	12000	9704	3.2	987	771	2.8
10.	Eastern Arc and Coastal Forest of Tanzania/Kenya	4000	1500	0.5	1019	121	0.4
11.	Western African Forests	9000	2250	0.8	1320	270	1.0
12.	Cape Floristic Province	8200	5682	1.9	562	53	0.2
13.	Succulent Karoo	4849	1940	0.6	472	45	0.2
14.	Mediterranean Basin	25000	13000	4.3	770	235	0.9
15.	Caucasus	6300	1600	0.5	632	59	0.2
16.	Sundaland	25000	15000	5.0	1800	701	2.6
17.	Wallacea	10000	1500	0.5	1142	529	1.9
18.	Philippines	7620	5832	1.9	1093	518	1.9
19.	Indo-Burma	13500	7000	2.3	2185	528	1.9
20.	South-Central China	12000	3500	1.2	1141	178	0.7
21.	Western Ghats/ Sri Lanka	4780	2180	0.7	1073	355	1.3
22.	South-western Australia	5469	4331	1.4	456	100	0.4
23.	New Caledonia	3332	2551	0.9	190	84	0.3
24.	New Zealand	2300	1865	0.6	217	136	0.5
25.	Polynesia/Micronesia	6557	3334	1.1	342	223	0.8
	Total		133.149	44.4		9645	35.3

Table 22.6. Comparison of Species Diversity in Two Sites with same Number of Individuals

Species	*Site A*	*Site B*
A	6	2
B	1	2
C	1	2
D	1	2
E	1	2
Total five species	10	10

The destruction of the world's tropical forests, which are disappearing at an alarming rate, is one of today's most urgent global environmental issues. A rich species diversity is slowly being lost for ever. Tropical forests are estimated to contain 50 to 90 per cent of the world's biodiversity. The report, based on Studies carried out by FAO and WCS, found the tropical forest & shrinking at the rate of 0.8 per cent each year (during 1980 to 1990). About 154 mha of tropical forests equivalent to almost three times the size of France have been destroyed. If the current rate of deforestation continues scientists estimate that roughly 5 to 10 per cent of tropical forest species may face extinction, within next one or two decades. The extent and loss of tropical forests in different ecosystems is shown in Tables 22.7, 22.8 and 22.9.

Table 22.7. Loss of Tropical Forests in Different Ecosystems

Types of forests	*During 1990*	*Annual decline (1981-90)*
Rain forests	713790	0.6%
Moist deciduous	519779	0.9%
Hills & mountain	201417	1.1 %
Dry deciduous	178579	0.9%
Very dry	59742	0.5%
Desert	8086	0.9%

Table 22.8. Loss of Forest in India during various projects or activities

Activity	*Area lost in ha*
River Valley Projects	401000
Agriculture	2433000
Roads and Communications	55000
Industries	125000
Miscellaneous	38800
Total	3052800

An average loss was 155000 ha per annum.

Table 22.9. Loss of Tropical Rain Forests at Global Level (1981-90)

Country	*Loss of forests*
Indonesia	22.6%
Brazil	23.6%
Malaysia	8.1%
Zaira	7.9%
Columbia	4.9%
Ali others	34.2%

Rainforests, the home to half of the world's life forms, continue to be destroyed at the rate of over 100,000 sq. km every year. This loss of biodiversity has immediate and long term effects on human survival. The majority of the world's population still depends on wild plant and animals for their daily food, medicine, housing and household material, agriculture, fodder, fuelwood and intellectual stimulation.

The loss is even more direct in the case of domesticated biodiversity. Traditional farmers of the world have developed an incredible variety of crops and livestock. This too has been eroded over the last few decades, as lakhs of traditional crop strains and hundreds of domesticated livestock breeds being replaced by a handful of laboratory-generated hybrids or by dominant cash crops.

The traditional diversity was bred to meet diverse human needs of nutrition, test, colour, ritual, smell and to resist drought, flood and pests. It provided several kinds of insurance against crop failure to the farmer. Modern hybrids, on the other hand, while substantially increasing the grain yield and monetary profits, have forced the farmers to look elsewhere for their other daily needs, especially fodder.

Estimating the precise rate of loss of biodiversity is a herculean task because no monitoring system, whatsoever, systematic is likely to match the huge diversity in its existing form. Thus, much of the basic information is lacking especially in the places like tropics and the hotspots. At least 10% of India's wild flora and fauna are threatened with many on the verge of extinction.

India's biodiversity is one of the most significant in the world. As many as 45,000 species of wild plants and over 77,000 of wild animals have been recorded, which comprise about 6.5 per cent of the world's known wildlife. An assessment of wildlife habitat loss in tropical Asia in 1986 showed that the country had only 6,15,095 sq. km out of its original wildlife habitat of 30,17,009 sq. km i.e., loss of about 20 per cent. In the last few decades, India has lost at least half of its forest polluted over 70 per cent of its water bodies, built on or cultivated much of its grassland, and degraded most of its coasts. Under such circumstances, none can say how many species have already lost.

The country has several problems such as over population, large number of cattle herds, growing demand for land, energy and water supply. Unplanned developmental works i.e., over exploitation of resources have made its living resources most vulnerable. Of the world's 12 top priority biodiversity hotspots, India has two within its boundaries. Over exploitation has not only resulted in shortage of various materials, but also left our biodiversity exposed to various ecological threats. Over emphasis on timber logging has affected many animal species. Faunal losses have been mainly attributed to over-exploitation of certain species for trading purposes, habitat alteration and destruction, and pollution of streams, lakes and coastal zones.

Listing of Threatened Biodiversity

Red Data Book is the name given to the book dealing with threatened plants or animals of any region. Many countries have prepared their own Red Data Book (e.g., Britain and New Zealand). On the global level, the IUCN published Red Data Book in two volumes. Its opposite is the Green Book, which lists rare plants growing in protected areas like botanical gardens. A mimeographed Green Book for India has been brought out for hundred rare plant species growing in the garden of Botanical Survey of India (BSI). The BSI has also compiled three volumes of Red Data Book having information of endangered plant species. The UNEP has compiled endangered species of the world under the title, 'Blue Book'. The IUCN has defined Red Data Categories which specify the state of extinction process.

Wildlife Status Categories

Vulnerable: These are the species whose population numbers are decreasing and are likely to become more severely threatened with time and in near future, they may enter into the category of endangered species, if unfavorable conditions in the environment continue to operate.

Endangered: The species with few individuals because of unfavorable environmental or human factors and that its natural regeneration is not able to keep pace with exploitation or destruction by natural and unnatural means. If the same factors continue to operate unabatedly then soon the species would become extinct, e.g., Indian Rhinoceros, Asiatic lion, Great Indian Bustard and white-winged duck of north-east India.

Rare: The species (or taxa) with small world population that are not at present endangered or vulnerable, but are at risk. Such species are usually localized within restricted geographical pockets or habitats or are thinly scattered over a more extensive range. Rare species have a population of less than 20,000 individuals. Some species are naturally rare and have never occurred in greater numbers, yet they are able to maintain these numbers. Other species have become rare through man's action or other unnatural forces.

Extinct: Species that are no longer known to exist in the wild but survive under captivity or cultivation. Generally, the term Extinct is used for the species that are no longer known to exist in the wild, e.g., the Cheetah and Lesser Indian Rhinoceros.

Threatened: It is a broader term that is used for species that fall into any of the vulnerable, rare and endangered categories.

Threatened Animals

Though the IUCN in 1988 listed 23 species of mammals as endangered or vulnerable to extinction, 75 species are totally protected as listed under schedule I of the Wildlife (Protection) Act 1972. Forests of the Western Ghat is famous for their endemic fauna and flora. The lion-tailed macaques, the Nilgiri Langur, the Malabar large-spotted civet, the Nilgiri tahr, etc., are some important endangered species. The Hoolock gibbon is the only ape found in the hilly forests of north-eastern India. Among the 19 primate species, 12 are endangered. The Manipur brown-antlered deer (Thamin), once distributed in some parts of the north-east, is perhaps the most threatened species. There are some 50 to 100 of them in the Keibul Lamjao on the bank of Logtak Lake sanctuary of Manipur.

About 427 endangered plant species have been listed by the BSI for the Red Data Book. This contributes about 20 per cent of the flouristic composition of higher plants of India. Some examples are given below :

Eastern Himalaya

Acer laevigatum, A. molle. Anglics anibigena, Buniurn not hum, Canim villosum. Pimpwella wallichi, Panax pseudogineseng, Calamus inermis, Phoenix nipicola, Lactuat cooperi, Berberis affinis, Engelharditia wallichiana, Coptis teeta, Aquillaria agallocha. Boehmeria terapensis.

North-eastern Himalaya

Hcraclcinn bunnanicum, Peucedanum sikkimensis, Pimpinellu evoluta, P. flaccida, Trachelospermum auritwu, Ilex embodies, I. khasiana, I. venulosa, Arnorphophallus bulbifer, A. sylvaticus, Dioscoria launfolia, D. arbiculata, Hopea shingkeng, Musa velutins, Hedychinm aitrantiacilin, H. gracilinium, H. dekiantim, H. gratum, H. grcetiii, IL liookcn, H. inarginatitm, H. rubrum.

Over exploitation is one of the major causes for the present endangered status of the Himalayan musk deer (Moschus moschiojerous). Human population has also caused depletion in the population of the Kashmir stag or Hangul. Indian wild ass and buffalo are also among the threatened mammals. The cheetah, and the Lesser Indian Rhinoceros have been extinct according to IUCN's Red Data Book. Other animals appear on the list are, Asiatic lion, Snow leopard, Swamp deer, elephant and tiger.

According to Dr. G. M. Oza, President of the Indian Society of Naturalists, there are atleast 30 mammals in India classified under the category of threatened species and our country tops in the list of countries with endangered mammalian species. After India, the US and China jointly occupy the second place with 29 mammals in both the countries on the threatened species list. Brazil, with 25 endangered species ranks third in the list. Further, in the oriental region, the country stands first in the threatened mammalian species. Indonesia with 20 threatened species ranks second and Nepal and Thailand, with 14 threatened mammals jointly occupy the third place. Both Burma and Pakistan have 13 threatened species of mammals and Sri Lanka with only four threatened species ranks fifth in the list.

There are 1175 species of the birds, including 42 endemics and 11 of these are considered threatened. N. J. Collar and P. Andrew in their book, Birds to Watch, have identified 70 species of threatened birds, 63 on the mainland, four on the

Andaman Islands and three on the Nicobar Islands. A number of threatened endemic birds are from north-east. The wood-pigeon of the Western Ghats, the white-winged wood duck of north-east India and the forest owlet of the Satpura hills are among the threatened birds. IUCN Red Data Book includes Bengal florican, cheer pheasant and Great Indian Bustard as most endangered bird species.

The reptile fauna comprises 435 species including 238 species of snakes, about 30 kinds of turtles and many crocodiles and lizards. The tropical forests of Western Ghats alone support some 20 species of snakes. In addition to their commercial exploitation for skin, the dwindling forest ecosystem also exposes these creatures to danger of extinction. The four monitor lizards-the common Bengal monitor, the water monitor, the desert monitor and the yellow monitor have been declared as protected. Marine turtles, freshwater tortoises and crocodiles have also been exploited. The green turtle, the loggerhead and the Olive Ridley are the most widely consumed species. All the four species of sea turtles and five of mud turtles have been declared protected. The gharial, the saltwater crocodile and the marsh crocodile (mugger) are exploited for their valuable skins. The three are included in IUCN list of threatened animals. But the population of crocodiles has been increased through breeding centers in Uttar Pradesh, Andhra Pradesh, Tamil Nadu, Orissa and West Bengal.

The amphibian fauna comprises 182 species, which includes salamanders, caecilians, frogs and toads. There is a high degree of endemism and out of 112 endemics, 84 occur in the Western Ghats and 20 in the north-east. Three of the Indian species, namely the Himalayan newt, Tylatotrition verrucosus, the Malabar tree-toad, Pedostibes tuberculosis and the Garo Hills tree-toad. Pedostibes kcinpi have been identified as rare endangered species by B. K. Tikader, in his 'Threatened Animals of India'. The first mentioned is the only species of salamander known in the country and is distributed in Sikkim, north-west Bengal, Arunachal Pradesh and Manipur. The

Malabar tree-toad was recently rediscovered in the Silent Valley.

The endangered species of mammals, amphibians, reptiles and birds are listed below.

Mammals (Number: 85)

- Andaman wild pig (Bug Scrofa and S. amanensis)
- Bharal (Ovis nnlntra) Binturong (Arctictis binturong)
- Bison or gaur or mithun (Bos gaums)
- Black buck (Antelope cervicapra)
- Blue whale (Balaenopteni misculus)
- Brown antlered deer or Thamin (Cervus eldi)
- Capped langur (Presbytis pileatus)
- Caracal (Felis caracal)
- Cetecean species (all)
- Cheetah (Acinonyx jubatus)
- Chinese Pangolin (Manis pentadactyla)
- Chinkara or Indian Gazelle (Gazella gazella bennetti)
- Chital or spotted deer. (Asix axis)
- Clouded leopard (Neofelis nebitlosa)
- Crab eating macaque (Maca'ca ints umbrosa)
- Desert cat (Felis libyca)
- Dugong (Dugong dugong)
- Fishing cat (Felis viverrina)
- Flying squirrels-all species of genera: Petaurista
- Eupetaurus, Belomys and Hylopetes
- Four horned antelope (Tetraceros quadricornis)
- Fresh water gangaitic dolphin (Platinista gangaitica)
- Gaint squirrels (Ralufa niacroura, R. indica. R. bicolor)
- Golden cat (Felis temmincki)

- Golden langur (Presbytis Seei)
- Gorals (Nemorhaedus goral, N. hodgsoni)
- Himalayan black bear (Selenarctos tibetanus)
- Himalayan brown bear (Ursus arctos)
- Himalayan crestless procupine (Hystrix hodgsoni)
- Himalayan ibex (Copra ibex)
- Himalayan tahr (Hemitragus jemlahicus)
- Hispid hare (Caprolagus hispidus)
- Hog badger (Arctonyx collaris)
- Hoolock gibbon (Hylobates hoolock)
- Humpbacked whale (Megaptera nectvangliae)
- Hyaena (Hyaena hyaena)
- Indian elephant (Elephas maximums)
- Indian lion (Panthera leo persica)
- Indian wild ass (Equus hemionus khur)
- Indian wolf (Cams lupus pallipes)
- Kashmir stag or hangul (Cervus elephus hangtil)
- Leaf Monkey (Presbytis melalophos)
- Leopard cat (Felis bengalcnsis)
- Leopard or panther (Panthera pardiis)
- Lesser or red panda (Wihirusfulgens)
- Lion tailed macaca (Macaca silenus)
- Loris (Loris tardigradus)
- Lynx (Felis lyns isabellinus)
- Malabar civet (Viverni megaspila)
- Malay or sun hear (I lelurctos malanus)
- Marbled cat (Felis marmorate)
- Markhor (Capra fulconeri)
- Mouse deer (Trugulus meminna)
- Musk deer (Moschus moschiferus)

- Nilgai (Boselaphus trgaocamelus)
- Nilgiri langur (Presbytis johni)
- Nilgiri fahr (Hemitragus hylocriux)
- Otters (Lutra lutra, L. perspicilata, Aonyx cynerea)
- Ovis amnion or nyan (Ovis ammon hodqsoni)
- Pallas cat (Felis mamus)
- Pangolin (Manis crassicaudata)
- Pig tailed macaque (Maciica memestrinu)
- Pygmy hog (Sus sylvanius)
- Ratel (Mellivora carpensis)
- Red fox (Vulpes vulpes)
- Rhinoceros (Rhinoceros unicornis)
- Rusty spotted cat (Felis rubiginosa)
- Sambar or barking deer (Cervus unicolor)
- Serow (Carpricornis sumatraensis)
- Sloth bear (Melursus ursinus)
- Slow loris (Nycticebus coucang)
- Small travencore flying squirrel (Pteromys fuscocapillus)
- Snow leopard (Panthera uncia)
- Spotted linsang (Prionodan pardicolor)
- Swamp deer or gond (Cervus duvauceli and other species)
- Takin or mishmitakin (Budorcas taxicolor)
- Tibetan antelope or Chiru (Panthelopus hodgesani)
- Tibetan fox (Vulpus ferrilatus)
- Tibetan gazelle (Procapra picticaudata)
- Tibetan wild ass (Epuus hemionus kiang
- Tiger (Panthera tigris)
- Urial or Shapru (Ovis vignei)

- Wild buffalo (Bubalus bubalis)
- Wild dog or dhole (Ciuon alpinus and other species)
- Wild yak (Bos grunnies)

Amphibians

- Himalayan newt or Salamander (Tryletotriton verrucosus)
- Malabar tree-toad (Pedostibes tuberculosus)
- Garo hills tree-toad (Pedostibes kempii)
- Viviparous toad (Nyctophyme tuberculosa)

Reptiles (Number: 20)

- Agra monitor lizard (Varanus qriseus)
- Atlantic ridley turtle or Kemps ridley turtle, (Lepidochelys kempii)
- Barred oval yellow monitor lizard (Varanus flavescens)
- Estuarine crocodile or salt water (Crocodylus porosus)
- Fresh water crocodile (Crocodylus palustris)
- Gharial (Gavialis gangeticus)
- Ganges soft shelled turtle (Trionyx gangeticus)
- Green sea turtle (Chelonia mydas)
- Hawksbill turtle (Eretmochelys imbricata inslacata)
- Indian egg eating snake (Elachistodon westermanni)
- Indian soft shelled turtle (Lissemys punctata punctata)
- Indian tent turtle (Kachuga testa facts)
- Large Bengal monitor lizard (Varanus Bengalensis)
- Leathery turtle (Dermochelys coracea)
- Loggerhead turtle (Caretta caretta)
- Olive back loggerhead turtle (Lepidochelys olivacea)
- Pythons (All species of genus python)

- Three keeled turtle (Geoemyda tricarinata)
- Tortoise (All member of the families : Testudinidae and trionovechidae)
- Water lizard (Varanus seawater)

Birds (Number: 38)

- Andaman teal (Assam gibberifrons albogularis)
- Assam bamboo Patridge (Bambusicola fytechii)
- Bazas (Avicedn gerdoni and A. leuphotes)
- Bengal florican (Eupodotis bengalensis)
- Black necked crane (Gnis nigricollis)
- Blood pheasants (Ithaginis cruentus tibetanus, I. cruentus kukseri)
- Brown headed gull (Lotus brunnicehalus)
- Cheer pheasant (Catreus Wallichii)
- Comb duck (Sarkidiornis Melanotos)
- Forest spotted owlet (Athens blewiti)
- Great Indian bustard (Chariotis nigriceps)
- Great Indian hornbill (Buceros bicornis)
- Hooded crane (Grus monacha)
- Hombills (Ptilolaemus tickelli, Access nipalensis, Rhyticeros undulates ticehursti)
- Hoabra bustard (Chlamydotis undulate)
- Humes bar backed pheasant (Syrmaticus humiae)
- Indian Pied hornbill (Anthracoceros malabaricus)
- Jerdons courser (Cursorius bitorquatus)
- Lammergeier (Gypaetus barbetus)
- Large falcons (Falco peregrinus, F. biarmicus, F. chicquera)
- Large whistling teal (Dendrocygna bicolor)
- Monal pheasants (Lophophorus impejanus)

- Mountain quail (Ophrysia superciliosa)
- Narcondom hornbill (Rhyticeros undulatus narcondami)
- Nicobar megapode (Megapodius freycinet)
- Nicobar pigeon (Caloenas nicobarica pelewensis)
- Peacock pheasant (Polyplectron bicalcaratum)
- Peafowl (Pavo cristatus)
- Pink headed duck (Rhodonessa caryophyllacea)
- Sdaters monal (Lophophourus sclateri)
- Siberian crane (Grus leucogeranus)
- Spur fowl (Gallaperdix spp.)
- Tibetan snow cock (Tetraogallus tibetanus)
- Tragopan pheasants (Tragopan melanocephalus, T. Blythii, T. Satyra, T. temminckii)
- White bellied sea eagle (Haliaeetus leucogaster)
- White eared pheasant (Croosoptilan crossoptilan)
- White spoon bill (Pistoled leucorodia)
- White winged wood duck (Cairina scutulata)

The status of many invertebrate species is not known. Lack of data is perhaps why Schedule I of the Wildlife (Protection) Act 1972 includes only a few of the species the Cocount crag of Andaman and Nicobar Islands, the Himalayan dragonfly and 130 species of butterflies. A number of species, beetles and butterflies are also becoming rare or vulnerable to extinction, either due to excessive exploitation or habitat destruction. Two important species of butterflies which are involved in the trade are, Kaiser-I-Hind, the swallow-tail butterfly found along the Eastern Himalayas, and the Bhutan Glory, distributed in Assam, Sikkim, Manipur and Nagaland. Besides the well-known sacred chank of India, Trochus and Turbo are the shell fish that deserve special mention and trade controls.

Threatened Plants

Plants of Ornamental Value

- *Aeridcs crispum Lindl,* (Orchidaceae) is an orchid with showy flowers; grows in certain regions of peninsular India.
- *Cymbidium aloiflium* Sw. (Orchidaceae) has not only attractive flowers but is reported to be medicinally useful and hence in great demand.
- *Diplomeris hirsuta* (Orchidaceae). The species is seriously threatened as it grows in parts of Darjeeling and Meghalaya only and is vulnerable to landslides and forest clearance.
- *Pabhiopedilum fairieyanum* (Lindl.). Pfitz. (Orchidaceae) or Lady's Slipper Orchid. A ground orchid which occurs in Arunachal Pradesh has been badly disturbed due to constant clearance and burning of forests for "Jhum cultivation". It is so rare that it was once labelled as the long lost orchid.
- *Qaphiopedilum drurui* (Orchidaceae) an orchid of medicinal value found in Kerala, has become very rare.
- *Rhododendron edgewortythii* Hk, f, (Ericaceae) is a showy epiphyte, reported to have been collected only on three occasions during the last one hundred years.
- *Symplocos cliengapae* Raiz. & Sahni (Symplocaceae) is an ornamental tree with fragrant flowers, confined to Nicobar Islands.

Plants of Medicinal Value

- *Acorus calamus* (Araceae) or 'Safed Bach'. The plant species which grew abundantly in Kashmir, Manipur and Nagaland, only a few years back, is now being imported to meet the medicinal demand of the country.
- *Atropa acuminata* Royle ex Lindl. (solanaceae) or 'Sag-angur'. This species is a source of some active

principles. The plant found to grow in Western Himalayas is rare now. Several species of this genus have also become rare.

- *Dioscorea deltodea* (Dioscoreaceae). The tubers of the plant species are used to cure rheumatic illness and for opthalmic problems. The species found to grow in the western Himalayas, Punjab and Kashmir is now facing extinction due to over exploitation.
- *Drosera* species (Droseraceac). A small insectivorous plant which is not only a botanical curiosity but also famous for its medicinal value. The population of the plant has become rare now due to over-exploitation.
- *Rauwolfia serpentina* Benth. Ex. Kurz (Apocyanaceae) or 'Sarpagandha'. The roots of the plants species arc highly valuable in medicine. The alkaloid, reserpine prepared from the roots has its value in treating high blood pressure. The species is nearly at the verge of extinction due to over-exploitation for commercial production of reserpine.
- *Saussurea lappa* Clarke (Asteraccae) or 'Kuth'. Its roots are highly valuable in medicine. The species found to grow in higher altitudes of the Himalayas has become rare.

Plants of Scientific Value

These plants exhibit some morphological interesting features and hence, are having, academic interest also.

- *Balanophora involurata* Hk.f. (Balanophoraceae) is a fiesry tuberous parasite on roots of trees, occurring in the Himalaya. It has been collected only on a few occasions so far.
- *Dischidia benagalensis* Coleb (Asclepiadaceae) is climber found in eastern Nepal and Sikkim. Leaves develop into pitchers which collect humus and debris and into these, special branched roots develop from the petiole for absorption.

- *Nepenthes khasiana* Hk. f. (Nepenthaceae) is an insectivorous pitcher plant. The leaf tips of the plant develop into curious pitcher like structures arched oval by lids. The pitchers are up to about 25 cm long. This plant has a special botanical curiosity. The population of this species, restricted to the khasi hills in Meghalaya, is not enough even to supply educational institutions of our country. The native habitat of the plant is now a gene sanctuary.
- *Sapria himalayana* Griff. (Rafflesiaceae) occurs as root pa rasite in Mishmi Hills of Arunachal Pradesh. Its flowers resemble the well known Reflesia flower and are up to about 35 cm in diameter.

Plants of Phytogeographic Importance

Most of the Indian endemic plant species show restricted distribution and are rare which need protection:

- *Ceropegia jainii* Ansari & Kulkarni (Asclepiadaceae) is an erect tuberous herb endemic to Ratnagiri in Maharashtra.
- *Frersa idica Dalz.* (Asclepiadaceae) grows in hilly regions of Maharashtra and has showy flowers.
- *Ghiphochola mysorensis* (Jain & Hem) Clayton (Poaceae/Graminae) is a grass endemic to Karnataka.
- *Helicanthes danser* (Loranthaceae) is a genus of parasitic plants, confined to the mountain of peninsular India.
- *Jainia nicobarica* Balak (Rubiaceae) is a plant endemic to Andaman & Nicobar Islands.
- *Manisuris divergens* (Hack.) Ktze (Poaceae Graminae) is a grass probably first collected in peninsular India in 1832. Its second collection was in 1963, after 13 decades. It is confined to Western Ghats.
- Willisia warm (Podostemaceae) is a genus of small herbs which grows on rocks in swift-flowing system of Western Ghats.

Trees of forestry Importance

While most trees of silvicultural importance are carefully looked after by the Forest Departments, some rare species do need protection:

- *Dysoxyhim malabaricum* Hedd. (Meliaceae) is a large tree that grows in Western Ghats. Its timber is ornamental.
- *Pterocarpus santalinus* Linn. F. (Fabaceae/Papiolionaceae) or 'Rakta chandan', commonly called Red Sandal, the valuable wavygrained timber. It yields insoluble blood-red dye, santalin, used for dyeing cloth, leather and wood. It grows in certain localised regions in Andhra Pradesh and its protection and expansion is very necessary.
- *Dccitssocarpus wallichianus* (Presa) de Lauben (Podocarpaceae) is the only conifer occurring wild in peninsular India.
- *Phyllostachys bambusoides* Sieb. & Zucc. (Poacea/Graminae) is common in China and Japan, but reported in India only recently. It is economically important.
- *Pinns gerardiana* wall (Pinaceae) or 'Chilgoza'. The fruits of the tree are used as a delicious dry fruit. Its distribution is restricted to a small region in Western Himalaya.
- *Santalum album* Linn. (Santalaceae) or 'Chandan'. Occurs commonly but its uncontrolled illicit exploitation and attack by spike disease want care and protection to the species.

Causes for the Loss of Biodiversity

a. Destruction of habitat: The natural habitat may be destroyed by man for his settlement, grazing grounds, agriculture, mining, industries, highway construction, drainage, dam, building, etc. As a consequence of this, the species must either "adapt to the changes, move elsewhere or may succumb to predation, starvation or disease and eventually die. In our country, several rare butterfly species are facing extinction due to habitat destruction in the Western Ghats. Of the 370 butterfly species available in the Ghats, up to 70 are at the brink of extinction.

b. Hunting: Commercially, wild animals are hunted for their products such as hides and skin, tusk, antlers, fur, meat, pharmaceuticals, perfumes, cosmetics and decoration purposes. For example, in Africa, in recent years 95% of the black Rhino population have been exterminated by poachers for their horn. Today, rhino horn fetches more than $ 15,000 in the pharmaceutical market. In the last seven years alone, over one-third Africa's elephants have been wiped out for some 3000 tonnes of ivory. In the international market, the cost of ivory is about $ 150 per kg, whereas in the Indian market, it varies between Rs. 2,000 to 2,500 per kg. Africa produces about 700 tonnes of raw ivory every year. The major buyers are Japan, Hong Kong, followed by the U.S., Germany and the U.K. The U.S. alone imports about $ 30 million worth of ivory annually, CITES regulations have to a great extent, reduced illegal trading and poaching of African Tuskers. Carving and sale of African ivory have been banned in India from April 2, 1992. In January 1987, the Central Government also banned the trade in Indian ivory (Agrawal, 1993.).

The scarlet macaw, once common throughout South America, has been eliminated from most of its range in Central America. Several species of spotted cats such as the Ocelot and Jaguar have been jeopardized by the demand for their fur. In 1962, nearly 70,000 whales were slaughtered. However, international trade in whale products is banned now.

In our country, rhino is hunted for its horns, tiger for bones and skin, musk deer for musk (medicinal value), elephant for ivory, Gharial and crocodile for their skin, and jackal for thriven fur trade in Kashmir. One of the most publicised commercial hunts is that of whale. The whalebone or 'baleen' is used to make combs and other products. According to WWF-Report, the spiny tailed lizard of the Indian desert, locally known as Sanda, is being hunted to extract the oil used for making 'Sanda ka tel' for its so-called aphrodisiac properties. The lizard is listed in schedule 11 of the Indian Wildlife Protection Act, 1972, which bans trade in the endangered species. But the WWF says, because of the huge demand from the quacks and doctors, the killing continued unabatedly.

CITES listed nine Indian animal species which have been severely depleted due to international trade. These are Fin whale (Balaciwptera physalus), Himalayan Musk deer (Moschus moschiferus), Green Turtle (Chelonia ydas), Hawksbill Turtle (Eretmochelya imbricata), Olive Ridley Turtle Dermochelys olivacea), Salt-water Crocodile (Crocodylus porosus), Desert Monitor Lizard (Varamtsgnsetis), Yellow Monitor Lizard (V. flavesoens), and Bengal Monitor Lizard (V. bengalensis).

Officials of TRAFFIC-India say poaching of the Indian tiger has risen because of increasing demand from south-east Asian countries and China, where pharmaceutical factories consume the bones of 100 tigers each year. Such demand has decimated the tiger population in China and brought the Russian tiger to the brink of extinction. As a result, in recent years much of the demand has been met by poachers in India. Smuggling of tiger bones and skins is a lucrative business. One kg of tiger bones fetches $ 90 in India and $ 300 in the international market. Hunting for sport is also a factor for loss of animal biodiversity.

c. Over-exploitation : This is one of the main cause of the loss of not only economic species, but also biological curiosities like the insectivorous and primitive species and other taxa needed for teaching or laboratory work (like Nepenthes, Gnetum, Psilotum, etc.). Commercial exploitation of biodiversity has invariably meant its overuse and eventual destruction. This was true in the case of whales, into plywood as of the whales, that were hunted for tallow, in the oceans.

Plants of medicinal value like Podophyllum hexandntrm, Coptis teeta, Aconitum, Discorea deltoidea, Rauwolfia serpentina, Qctphiopedilum druri, etc., and horticultural plants like orchids and rhododendrons come under the over-exploitation. For instance, excessively harvesting of marine organisms such as fish, molluscs, sea-cows and sea-turtles has resulted in reduction of these animals.

d. Collection for Zoo and Research: Animals and plants are collected throughout the world for zoos and biological laboratories for study and research in science and medicine. For example, primates such as monkeys and chimpanzees are sacrificed for research as they have anatomical, genetic and physiological similarities to human beings.

e. Introduction of Allan Species: Native species are subjected to competition for food and space due to the introduction of exotic species. For example, introduction of goats and rabbits in the Pacific and Indian regions has resulted in destruction of habitats of several plants, birds and reptiles.

f. Control of Pests and Predators: Predator and pest control measures, generally kill predators that are a component of balanced ecosystem and may also indiscriminately poison non-target species.

g. Pollution: Pollution alters the natural habitat. Water pollution especially injurious to the biotic components of estuary and coastal ecosystem. Toxic wastes entering the water bodies disturb the food chain, and so to the aquatic ecosystems. Insecticides, pesticides, sulphur and nitrogen oxides, acid rain, ozone depletion and global warming too, affect adversely the plant and animal species. The impact of coastal pollution is also very important. It is seen that coral reefs are being threatened by pollution from industrialization, oil transport and offshore mining along the coastal areas.

Noise pollution is also the cause of wildlife extinction. This has been evidenced by the Canadian Wildlife Protection Fund. According to a study Arctic Whales are seen on the verge of extinction as a result of increasing noise of ships, particularly ice breakers and tankers.

h. Deforestation: One of the main causes for the loss of biodiversity is population explosion and resultant deforestation. Deforestation mainly results from population settlement, shifting cultivation, development projects, demand for fuelwood, demand of wood for industry and other commercial purpose. Figure 22.2 depicts the increasing production of forest based materials. The report, based on studies carried out by the FAO and WCU, found that tropical forest areas are shrinking at the rate of 0.8 per cent each year. Between 1980-90, 154 mha of tropical forests equivalent to almost 3 times the size of France have converted to other uses. The forest area lost in country is shown in Table 22.8. This represents an average annual loss of 1,55,000 ha.

Fig. 22.1. Loss of Biodiversity due to deforestation

In India, the rate of deforestation is 13,000 sq. km annually. If this rate of deforestation continues, one can imagine the ultimate fate of our forest and biological richness. It is presumed that in coming years, the global loss of biodiversity from deforestation alone would be 100 species everyday.

i. Other Factors: Other ecological factors that may also contribute to the extinction of plant and animal diversity are as follows :

1. Distribution range. The smaller the range of distribution, the greater the threat of extinction.

2. Degree of specialization. The more specialised an organism is, the more vulnerable it is to extinction.
3. Position of the organism in the food chain. The higher the organism in food chain, the more susceptible it becomes.
4. Reproductive rate. Large organisms tend to produce fewer offsprings at widely intervals.

Fig. 22.2. Increase in forest products due to afforestation

Socio-economic and Political Causes

The socio-economic and political causes of biodiversity loss vary from region to region. In recent times, they can be linked to governmental and international support for industry, forestry, agriculture and energy programmes. The promotion of subsidy for big mono-cultures of soyabeans did not make place to feed hungry Brazilians, but to feed fat cows here in the common market, cows that are producing what even the technocrats of the European market call seas of milk and mountains of butter. It was this policy of promoting cash crops in the south that destroyed one of the most well adapted, healthy and ecologically safe peasant cultures that it was in the south of Brazil.

In south-East Asia, it is the tropical timber industry that is cutting down the last primary rainforests, against the will of local people (Nectoux and Kuroda, 1989; Colchester, 1989; Hong, 1986). As in Brazil, the traditional practices and rights of local communities are violated and the destruction of resources for the benefit of the few is justified in the name of national development and the generation of foreign exchange. The role of western culture, i.e. use and throwaway culture that causes the liquidation of natural resources for short-term profit, is also one of the important causes of deforestation, and thereby the loss of biodiversity in industrialized countries.

The introduction of exotic species of plants and animals causes major disturbances to ecosystems because the checks and balances that hold the equilibrium in the native ecosystem do not apply to the introduced species. In developing countries like India, the development policies and projects have rarely been sensitive to the need for biodiversity conservation, and that of the local communities. The Government's failure to remove poverty and curb middle-class consumerism has led to conditions in which sensible natural resources management assumes low priority.

Biotechnology Biodiversity and IPR

In the last twenty years, all over the world, there has been a revolution in the field of biotechnology. New discoveries and inventions in the area of isolation and manipulation of genes, better understanding of biological molecules and the advent of recombinant DNA technique have increased our knowledge. Biotechnologists all over the world have made efforts to create transgenic crops which will with-stand the pests as also have enough resistance to withstand environmental stress. In fact, biotechnology is inherently knowledge-intensive and having strong infrastructure would lead to value area of agriculture, animal husbandry, fisheries, forestry and medicine.

Biotechnology is not a miracle solution to the problem of biodiversity crisis. Rather, the use of biotechnology in the production of uniformity in plants and animals has threatened not only the life forms but also rendered entire community or ecosystem unstable. Further, indiscriminate and unregulated uses of genetically modified organisms pose a threat to mankind. In fact, uniformity (or homogeneity) in life forms accelerates the loss of biodiversity. The institutional structure that controls the biotechnology, therefore, should not over shadow those institutions that deal with conservation of Biodiversity.

Productivity and Diversity

Productivity goes against diversity as it creates imperative for uniformity and homogenization. This has generated the paradoxical situation in which modern plant improvement is based of the logic of uniformity and homogenization. Green revolution, for instance is based on high productivity and low biodiversity. There is no need to combine high productivity and high genetic diversity to enhance yield as well as to provide insulation against environmental stress and pollutants. Over the last few decades lakhs of traditional crop strains and hundreds of domesticated livestock breeds have been replaced by a handful of laboratory-generated hybrids or dominant cash crops. Similarly, forestry schemes introduced mono-cultures of commercial species like teak, eucalyptus and bamboo, and pushed into extinction the diversity of local species. Agriculture modernization fisheries, commercial forestry and animal husbandry thus produce uniform crops and domesticated livestock's and destroy the diversity of local species which fulfil local needs. Such a strategy of productivity increase based on the logic of destruction of biodiversity is no longer desirable as it will ultimately lead to loss of biodiversity.

Mono-cultures are ecologically unstable. Being genetically uniform, they invite diseases and pests, also vulnerable to environmental stress and pollutants. The technology for breeding high yielding varieties, indeed, a technology which breeds uniformity and at the same time threatens the biodiversity conservation and sustainability.

Politics of Patents and Intellectual Property Rights

Biotechnological processes use life forms of derivatives therefore to make or modify products or processes for specific use. Under Intellectual Property Rights (IPRs), transformed micro-organisms, plants and animals can be patented and become exclusive private property. The north has always used third world germplasm as a freely available resource and modified it. The issue of patent protection for modified life forms raises a number of unresolved political question about the ownership and control of the genetic resource. By simply manipulating the life forms one does not acquires the patent or property right, because the modified life forms do not arise from nothing but from existing life-forms which belonging to others. Also biotechnology does not create new genes, but merely relocates genes already existing in the organism. The advanced capitalist nations wish to retain free access to the developing world's storehouse of genetic diversity, while the south like to have the proprietary varieties of the worth's industry declared a similarly public good. The north, however, resists this democracy. US has freely taken the biological diversity of third world to earn millions of dollars, none of which have been shared with third world countries, the original owners of the biological resource. For instance, an American industry earned $ 8 million a year in 1962 simply by increasing the soluble solid contents of a wild tomato variety, Lycopersicum chomrelewski taken from Peru. None of these profits was shared with Peru, the original contributor of the genetic material. The Convention on Biological Diversity is also not clear on this. Industrialized countries, particularly the US interpreted key clauses of the treaty in a manner that would protect the interest of its own biotechnology industries. This is a clear setback to the developing countries, who stand to lose the benefits due to them.

In absence of a proper biotech base, a developing country cannot match an industrial country although the former may be far richer in Biodiversity. However, the Convention on Biodiversity helped to place the subject matter of technology transfer and IPRs on the top of the agenda of policy and decision makers. Furthermore, access to genetic resources and transfer of technology is treated on the same plan. On the issue of IPRs, the basic requirement of the Dunkel proposals is that inventions in all branches of technology shall be patentable, whether products or processes, if they meet the three tests of being new, involving innovative steps and being capable of industrial applications. It has also been provided that micro-organisms will be patentable. In respect of plant varieties, there is a separate obligation to provide them protection by patents by an effective sui-generis system, or by a combination of the two. Sui-generis protection implies a system different from other categories of intellectual property protection (such as patents) and is a class by itself. Dunkel text, thus does not compel to patent seeds (i.e. plant varieties). So far we are concerned, seeds are also not patentable in India today, and we do not have any intention of changing this system. However, we will adopt our own system for the protection of plant varieties under which we may provide certificates for plant breeders right. The farmer's rights include their using the seeds for their own needs or for exchange in the village community according to their traditional custom. Since farmer's right will be fully safeguarded under system of protecting the plant breeders right, there is no truth in the allegation that the farmers will not be able to retain the seeds for their own use and that they will have to buy seeds every year from multinational companies. Furthermore, India is not in favour of the patenting naturally occurring life forms/ germplasm.

The extension of IPRs to plant varieties either in the form of patents or in the form of plant breeders rights is bound to result in increased prices of seeds, greater domination of agriculture by multinational companies and slower diffusion of new varieties. These would be in sharp contrast to the experience of the green revolution where the new varieties of seeds evolved by the government institutions percolated down to the fields in a short span of time with very little cost to the individual farmer. Our farmers will have to face great hardship due to the new regime. India calls for the removal of distortions in the IPRs regimes in areas related to prior existence of knowledge in nutrition. Mr. Kamal Nath, former Environment and Forest Minster, referred to the harsh fact that IPRs regimes in these areas resulted in a virtual denial of benefit flows of financial return and markets to those very communities who had by their sustainable life styles preserved these systems and the natural resources on which they were based. It is said that the market value of allopathic medicines derived from plants used in traditional remedies was calculated to be over $ 43 billion annually. Less than 0.01 per cent of profits had gone to the indigenous people who had led the researchers to them.

The south had not merely preserved much of its precious biogenetic resources, but also the knowledge and practices about their optimum and sustainable utilisation. Access to these resources have to be regulated and careful exercise, in keeping with the objectives of the convention and with due compensation to such people who have preserved their resources.

How to recognize and measure the value of indigenous knowledge, is one of the basic problems in deciding the compensation and for protection of farmers IPRs. As a result of the persistent North-South split, the CSD could able to move forward on this contentious issue. However, the decision of knowledge, innovations and practices of indigenous and local communities is an important step in the direction of the protection of traditional knowledge and practices of the indigenous and local communities relevant to conservation and sustainable use of biological diversity.

Biosafety Protocol

India has demanded a clear comprehensive and equally binding international protocol on biosafety under the convention on biodiversity. Addressing the first meeting of parties to the Convention on Biodiversity, at Nassau, Bahamas, on December 8, 1994, Mr. Kamal Nath demanded immediate and adequate safeguards against hasty experimentation and use of genetically modified organisms, since these have unimaginable repercussions. He feared that the developed world could become a play ground for experimentation with such genetically modified organisms and it could be checked through a legal binding agreement.

Fig. 22.3 Biosafety Protocol

7

BIODIVERSITY IN WESTERN GHATS

The biological diversity of Indian subcontinent is one of the richest in the world, owing to its vast geographic area and climate, and the juxtaposition of several biogeographical regions. Of the 12 megabio-diversity regions of the world, India is one. The megadiversity regions are identified on the basis of overall species diversity. The groups of animals viz., vertebrates, swallow tails, butterflies and higher plants are used to identify the mega-bio-diversity regions. The mega-bio-diversity regions include, Mexico, Columbia, Ecuador, Peru, Brazil, Zaire, Medagascar, China, India, Malaysia, Indonesia and Australia.

The greatest density of biological diversity is found in the Himalayas, the Andaman and Nicobar Islands and the Western Ghats. Out of the 25 hot spots identified in the world, India has two. They are Eastern Himalayas and Western Ghats. The Western Ghats are a continuous stretch of mountains, except for Palghat gap, presenting rich and varied types of flora and fauna. They have different vegetation types: scrub jungles, grasslands at low altitudes, dry and moist-deciduous forests, montane grasslands, sholas and precious tropical evergreen and semi-evergreen forests. The hot spots are identified based on the presence of greatest number of endemic species. At global level, these areas are of high conservation priority because if unique species are lost, they can never be replaced. This particular criteria was used by Myres (2000) to identify the 25 hot spots on the earth.

Of the estimated 15,000 flowering plant species in India, about 4500 occur in the Western Ghat region. Gramineae (120 genera and 400 species), Leguminosae (85 genera and 325 species), Orchidaceae (60 genera and 250 species), Acanthaceae (45 genera and 65 species), Cyperaceae (21 genera and 160 species), Euphorbiaceae (55 genera and 155 species), Compositae (57 genera and 130 species), Labiatae (25 genera and 115 species), Rubiaceae (45 genera and 100 species) and Asclepiadaceae (30 genera and 90 species) are the ten dominant families in the region. There is also a great diversity of traditional and medicinal plant species. The Western Ghats are home for many endemic, rare, and endangered species, as well as scores of economically important species and wild relatives of cultivated plants. The southern areas in particular have the richest gene pool of the region. Of the 4500 species in the Western Ghats, 1720 are endemic. Nearly one-third are rare or threatened, and several are believed to be extinct (e.g., *Dalbergia travancorica and Vanda wightii*). The endemic species are those species which are restricted in distribution. They have been evolved to adapt the special conditions in that area. If their habitat is destroyed, for instance, by deforestation they can easily become extinct.

Fifty four genera of flowering plants are endemic to the region. Of these, 42 are monotypic and they represent only one species. An example is *Blefaristemma membranifolia.* Eighty endemic species of trees and shrubs in the Western Ghats are confined only to evergreen forests and are not found in other types of forests. Several woody genera of plants found here e.g., *Syzygiitni* (18 endemic species) and *Litsea* (fourteen endemic species). Eighty four species of endemic orchids reported in this area. The south Indian species of Ladies slipper orchid, *Paphiopeddi druryi* is an endangered species found only in Agastyamalai hills. The following area in the Western Ghats are rich in endemism.

1. Agastyamalai hills	150spp
2. Nilgiri hills	93spp
3. Anaimudi	30spp
4. Palani hills	18spp
5. Anamalai	13spp

The following groups are endemic to Western Ghats

1. Gramineae	10 genera/150 species
2. Orchidaceae	2 genera/84 species
3. PnImae	12 species
4. Impatiens	71 species
5. Piper	9 species
6. Dalbergia	9 species
7. Oclilandra	8 species

The following species are most endangered or believed extinct or in the verge of extinction.

Trees

1. *Hopca jacobii*	Dipterocarpaceae
2. *Nothopegia aureo-fulva*	Anacardiaceae
3. *Cynometra beddomci*	Leguminosae
4. *Syzygium palghatense*	Myrtaceae
5. *Actinodaphne bourncae*	Lauraceae

Herbaceous Species

1. *Haplothismia exannitlata*	Burmaniaceae
2. *Hitbbardia heptancnron*	Gramineae
3. *linpatiense anaimudica*	Balsaminaceae
4. *Arisaema articulata*	Araceae
5. *Anoectochilus rotitndifolius*	Orchidaceae

The forests of Western Ghats are the veritable store houses of economically important species. Trees that yield timber and firewood include Dalbergia spp., Dipterocarpus spp. and Hopca spp.

Food Yielding

1. Edible fruits	*Artocarpus hirsiitus, Garciania India*
2. Edible nuts	*Stcrcidia guttata, Terminalia bellarica*
3. Spices and Condiments	*Garcinia gwnmigatta, Piper nigrnni*
4. Gums and resins	*Canariwn strictum Valeria indiai*
5. Tans and dyes	*Garcinia niorclla Terminalia spp.*
6. Essential oils	*Gymbopogon spp., Mesuu ferrca*
7. Medicinal	*Discorea spp., Rauwolfia serpentina*

The Western Ghat is also rich in wild relatives of plants namely rice, finger millet, pearl millets, sugarcane, pepper, turmeric, ginger and nutmeg. Western Ghats are the rich gene pool of wild relatives of many crop plants. These wild relatives play a vital role as gene sources in the plant breeding programmes. We need a reservoir of wild species to prevent the genetic erosion in our cultivated crops. This erosion can occur as farmer to grow a variety of species, instead he plants only a few high yielding varieties of rice and millets. The wild species *Sacchanmi spontaneiun* has been extensively used in breeding superior varieties of sugarcane *(Sacchanmi officnarum)*.

This part is also rich in faunal diversity and is the true home of the Indian fauna, however many species are now endangered or become most rare. Seven of the 63 mammals found in the Western Ghats are endemic. In addition there are 91 species of reptiles and 84 endemic species of amphibians found in the Western Ghats. The details about the faunal groups are described separately. A list of top storey, middle storey and understorey plant species found in the Bhadra Wildlife Sanctuary are showed in Table 23.0.

Table 23.0. Floral Diversity Recorded in the Bhadra Wildlife Sanctuary (Anon, 1995: Jain, 1996 : Raghavendra, 1997: Gina, 2000)

Key: TVS = important timber yielding spp.; FYS = important fruit yielding spp.; EIS = other economically important spp., Fr=fodder plants; Ex = exotic introduced species; M = medicinal properties; Fd=some part eaten

Botanical Name 1	*English Name* 2	*Local Name* 3	*Family* 4	*USES* 5
Top Storey				
Adina cordifolia	Kadam,	Ethega;	Rubiaceae	TVS, M, Fr
(Anthocephalus cadamba)	Kadamba	Kadawala		
Albizzia adoratissima		Biwara		
Albizzia lebbek		Siris		
Albizzia procera		Bellatte		
Alseodaphine semicarpifolia		Mashe		TVS
Alston/a scholaris	Devil's Tree	Hale; Maddale;	Apocynaceae	M, TVS, EIS
	(Chhatim) Jackfruit	Saptaparna	Moraceae	
		Halasu; Jaca		
Artocarus heterophyllus				
(A. integrifolia)				
Artocarpus hirsuta		Hebbalasu	FYS	
Ariocarpus lakoocha		Wate		
Bombax malabaricum	Indian Silk- Cotton Tree	Simul; Buruga	Bombacaceae	TYS, M, Fd, Fr
Madhuca indica	Mahwa Tree	Hippe	Sapotaceae	TYS, FD, EIS, FYS
(Bassia lateifolia)				*Bears eat flowers*

Calopyllum tomentosum (C. inophyllum)	*Alexandrian* Laurel??	*Surhonne*	*Guttiferae*	*M, TYS*
Canarium *strictum*		Kaldhupa		
Casuarina equisetifolia	*Beefwood Tree*	*Survey*	*Casuarinaceaea*	*EIS (fuel)*
Cinnamomum zeylanicum	*Cinnamon*	*Lavangpatti, Dalchinni*	*Lauraceae*	*M*
Dendrocalamus giganteus	*Bamboo*			
Elaeocarpus tuberculatus				
Eucalyptus ternicornis	*Eucalyptus*	*Nigiri*		*EIS*
Grewia tilaefolia		Thadasal		TYS
Gmelina arborea		Shivane		
Hopea *parviflora*		*Kiral Bhogi*		*EIS*
Lagerstroemia lanceolata		*Nandi*	*Lythraceae*	*TYS*
Lagerstroemia flosreginae (L, speciosa)	*The Queen's Flower, or Pride of India*	*Hole-dasavala Challa, Hloedachalla*	*Lythraceae*	*TYS, M, FYS*
Lannea coromandalica		*Godda*		
Mangifera indica	*Mango Tree*	*Mavu*	*Anacardiaceae*	*FYS, supports Many orchild spp.*
Machilus marcranfha		Gul mavu		
Mimucylon edule		*Anjan*		
Mitragyna parvifolia		Kadivala.		
Mimusops elengi		Moulsari; Ran		
Mesua ferrea	Ceylon Iron	Naka, Nagkesar,	Guttiferae	

	wood tree	Nagasamplige		
Mesua nagasarium			Guttiferae	
Melia azelarach (Azadirachta indica)	*Persian Lilac, Bead-Tree??*	*Nim, Hebevu*	*Meliaceae*	*M, TYS, EIS*
Michelia champeaca		*Sampige*		*EIS*
Michelia nilagirica				
Pterocarpus marsupium	*Indian Kino*	*Honne*	*Fabaceae*	*M, TYS*
Stereospermum chelenoidis		*Padri*		
Tectona grandis	*Teak*	*Taiga, Saguvani*	*Verbenaceae*	*TYS*
Terminalia bellirica	*Belleric Mycrobalan Mycrobalan*	*Taremara; Tare*	*Combretaceae*	*M, TYS*
Terminate tomentosa		*Mathi*	*Combretaceae*	*TYS*
Tertninalia paniculata		*Kindal*	*Combretaceae*	*TYS*
Terminalia arjuna	*The Arjuna*	*Bilimatti; Hural; Holematti*	*Combretaceae*	*M, EIS-tanning*
Terminalia chebula	*Chebulic Mycrobalan*	*Alalekayi; Alale*	*Combretaceae*	*M*
Tetrameles nudiflora		*kadubende*		
Thespesia populuea	*The Bhendi tree*	*Arasi, huvarsi*	*Malvacceae*	*TYS, M*
Valeria indica				
Vitex altissima	*Bharanige; Naviladi*			*EIS*

Middle storey

Acacia arabica		*Caesalpinaceae*		
Allophylus cobbe				
Anogerissus latifolia		*Dindage, Dindal*		*EIS, TYS*
Antidesma diandrum				
Aporosa lindleyana		*Challe*		
Butea monosperma	*The Flame of the Forest*	*Mutluga, Brahma uriksha*	*Papilionaceae*	*M; TYS, EIS*
Bauhinia purpurea	*Purple Bauhinia,*	*Basavanapadu*	*Laguminosae* (Caesalpiniaceae)	*Fr*
Bridelia retusa	*Moutain Ebony*			
Buchnania tanzan		*Badamara*		
Bischofia javanica		*Murkal*		
Careya arborea		*Kavalu*		
Cassia fistula	*Indian Laburnum*	*Kakkei, Kakkegida*	Caesalpiniaceae	M;TYS
Cassia siamea		*Seeme thandadi;* Simatangedu	*Caesalpiniaceae*	
Cordia maclodie		*Hadga*	*Baraginaceae*	
Cordia myxa		*Challe*	*Baraginaceae*	
Caryota urens	*Sago palm*	*Baine; Bagani*	*Palmae*	
Dillenia pentagyana		*Kanagal*	*Dilleniaceae*	
Diospyros monlana		*Jagalaganti*		
Dalbergia latifolia Rosezvood	*Bombay*	*Biti; Beete*	*Papilionaceae*	*TYS-one of most valuable*
Evodia lunuenkanda		*Kambli*		
Phyllanthus emblica (Embilica officianlis)	*Embli* Mycrobalan;	*Nellikayi; Nelli*	*Euphorbiaceae*	*FYS; M; EIS;*

	Indian Gooseberry			
Fagara badruntga		*jumminakai*		
Ficus bengalensis	*Banyan Tree*	*Ala, Atti, Aladamara; Athimara*	*Moraceae*	
Flacourtia monlana		*Mullu smapige*	*Flacourtiaceae*	
F. indica				
Gardenia ummifere		*Bikke*		
Gaicinia morella		*Murygana huli*		
Gmelina arborea				
Grewia hisuta				
Hopea wightna		*Haiga*		
Hydnocarpus laurifolia		*Neeruhanchi*	*Flacourtiaceae*	*M*
Holigarna		*Holegar*		
Kydia calycina		Bende		
Lagerstromemia parviflora		Chennangi	Lythraceae	
Mallotus philippenensis	Kamela (1)	Kunkumadamara	Euphorbiaceae	M;EIA;Fr
Microtropis latifolia				
Myristica fragrans	Nutmeg	Rampatrea	Myristicaceae	M;EIA
Olea dioca				
Pongamia glabra	The Pongam	Honge	Leguminosea	*M*
Schleichera olesoa	Gum Lac Tree	Kendale	Sapindaceae	TYS; EIA; FYS
Strychnos nuxvomica		Kasaraka		
Saraca asoca (S. indica)	Ashoka (1)	Asoka	Leguminosea	*M*
Salix tetrasperma	Willow			
Sapindus emarginatus		Antuvala		

Sapindus lauhfolius				
Santalum album	Sandalwood	Shrigandha	Santalaceae	M;EIA
Semicarpus anacardium		Kadugeru		
Spondias accuminaa		Kadamte		
Sterculia guttata			Sterculiaceae	
Sterculia villosa		Sahvige	Sterculiaceae	M
Stereospermum personatun				
Syzigium cumini (*Eugenia jambolana*)	The Jamun Tree; Java Plum	Nirale; Neralu	Myrtaceae	Fd;M
Tamarindus indica	Tamarind Tree	Hunasemara; Hunase	Leguminosae	Fd; M; TYS; EIA
Trema orientails	Charocal Tree			
Wrightia tincforia.		Hale		
Xylia xylocarpa		Jambe		
Under Storey				
Acacia intia	Acacia	Kadusige	Leguminosae (Mimosoideae)	
Acacia consinna		Seega		
Aeschynomene indica			Papilonaceae	
Agrostistachys indica				
Asclepias curassavica			Asclepidaceae	
Asystasia gangetifia			Santalaceae	
Bambusa arundinacea	Bamboo		Bambusaceae	
Bambusa bamboos	DowgaBamboo	Hebbiduru	Bambusaceae	
Didens pilosa			Asteraceae	

Borreria hispida			Rubiaceae	
Borrefia stricta			Rubiaceae	
Calycopteris floribanda		Neerballi		
Canscora diffiusa			Gentianaceae	
Carissa carandas		Karandakai		
Cassia auriculata			Caesalpinaceae	
Coccinia indica			Cucurbitaceae	
Corchorus capsularis			Tiliaceae	
Crotolaria laburnifolia			Papilonaceae	
Curcuma spp.		Arasina		
Cymbopoqan spp.	Lemon Grass		Poaceae	M
Cynoglossum			*Boraginaceae*	
Dinticulatum				
Dendrocalamus stnctus	Bamboo		Papolonaceae	Rare plant
Desmodium triquetrum				
Entada scandens		Ganpeballi		
Ficus hispida		Kadathimara.	Urticacaea	
Flemmgia strobilifera			Papilonaceae	
Glycosmis pentaphylla				
Holarrhena antidysentrica	Kurchi (1)	Kodacha;	Apocynaceae	M; Afforestation of poor soils-pioneer sp after clearance

Helecteres isora		Kowri		
Hemidesmus indicus	Indian	Sogadebera	Periplocaceae	*M*
Icnocarpus fmitescens	Sarsaparilla			
Imperata arundalaracea			Apocynaceae	
Ixora parvillora	Elephant Grass			
Jasminum spp.		Goravi	Rubiaceae	
Jasttcia simple		Kasumallige		
Lantana camara	Lantana	Santalaceae		
Leea sambusina		Nurchi		
Lucas aspera		Thumbe		
Lettsomia thomsoni		Uganiharnbu		
Malva sylvestris			Malvaceae	
Maytenus emarginata			Celastraceae	
Mimosa pudica		Nichige mullu	Mimosa pudica	
Mollugo oppositifolia			Alizoacaea	
Momordica dioica			Cucurbitaceae	
Murraya koeingie		Karibevu		
Ocimum sanctum	Sacred Basil;	Vishnu Tulasi	Lamiaceae	M
		Holy Basil		
Oldenlandia dichotoma			Rubiaceae	
Pavonia zeylancia			Malvaceae	
Peristrophe bicalyculate			Santalaceae	
Piper longum	Long Pepper	Pipal	Piperaceae	M
Plumbago zevlanica			Plumbaginaceae	
Randia dumatorum		Kare		

Randia uliginosa		Doddakare		
Reeds		Vate		
Rhyachosia beddomei			Pepilonaceae	
Sida acuta			Malvaceae	M
Sida cord/folia		Hiretutti	Malvaceae	M
Sida veronicaefolia			Malvaceae	
Solarium nigrum	Black Night-Sade Kari	Kachigida,	Solanaceae	M
Solarium Sisymbrifolium			Solanaceae	
Stachytorpeta indica			Verbenaceae	
Strobilanthus spp.		Kurungi		
Tephrosia purpurea			Papilonaceae	
Tridax procumbens			Asteraceae	
Triumfetta rhomboidea			Tihceae	
TiOnospora cord/folia	Tinospora	Amritaballi	Menispermaceae	*M*
Vicoa indica			Asteraceae	
Vitex negundo	Lakki			
Waltheria indica			Sterculiaceae	
Zizyphus ocnoplea	Bore; paragi			
Zizyphus zuzuba	Bore; yelachi			

Climbers: Argyreia cuneata Cardisopermum halicacabum, Cayratia pedate, Coccinia indica, Hemidesmus indicus, Ipomeae campanulata, Ipomeae piltata, Lantana camara and Momordica dioca.

Grassland species: Themeda tremula, Habenaria lorgicomiculata, Cymbopogaon caesisu, Heteropogon contortus, Eulalia trispicata, Jansenella grffithiana etc.

Fin Fish

Aquatic Resources of Karnataka

Karnataka is rich in fish resources. The fisheries, as such, have a vital economic role to play in the upliftment of community life of rural areas through augmenting production of nutritious protein rich food with immense potential for creation of rural employment. The resources are as follows :

Rivers : The total length of the riverine resources of the state is around 5 813 km.

Reservoirs : There are 73 reservoirs commanding a total water spread area of 2.25 lakh ha. and about 3000 km of irrigation canals.

Tanks : The state has 25,712 major and minor tanks having a total water spread area of 2.93 lakh ha. These are, in effect, seasonal, long seasonal and perennial in nature.

Marine : The state has a coast line of 300 km. having 27,000 sq. km. continental shelf and 87,000 sq. km. exclusive economic one.

Estuarine: There are 26 estuaries having a total water spread area of 7214 ha.

Brackish water : The state has a brackishwater area of 8000 ha.

Status of fin fishes in Karnataka

Inland Waters

Depending on their ecology, fishes are variously adapted to environmental conditions in the aquatic ecosystem. Some have excellent hydro-dynamic characteristics enabling them to swim at high speed while others lead a relatively immobile mode of life acquiring a number of adaptations in the course of evolution that are unrelated to hydrodynamics but that ensures normal development and survival of the species. The wide differences in drainage and sub-drainage or rivers across which the impoundments are situated, zoo-geographically these have determined the fish species available in each. There are, in all 175 fish species recorded in the state. Commercially important fish species have been enumerated in Table 23.1.

In Krishna river basin, 101 fish species belonging to 19 families and 5 orders were recorded which also includes the Indian Gangetic carps and the exotic fish. Majority of the fish species in this riverine system are endemic and, of this 20 species are under "threat of extinction". The Cauvery river basin has 96 fish species under 23 families and 8 orders which, as in the previous case, also includes the Gangetic major carps and the exotic fish species. The carps, *Puntius carnaticus, Hypselobarbus dubius* and the non predatory catfish *Pangasius pangaslus* available earlier, are now no more recorded in the riverine zone of the State. In addition, 15 more fish species are at the verge of total elimination. Sixty fish species belonging to 13 families and 4 orders were recorded in the riverine stretch of the Godavari basin which also includes the introduced Gangetic and the exotic carp species. Of these, *Thynnichthys sandkhol* has become "extinct", in addition, 15 more fish species are under the "threat" of elimination.

Table 23.1. Commercially Important Freshwater Fish of Karnataka

1. *Notopterus notopterus* (Pallas)
2. *Anguilla bengalensis bengalensis* (Gray)
3. *Anguilla bicolor bicolor* McClelland
4. Catla *catla* (Hamilton-Buchanan)
5. *Cirrhinus cirrhinus* (Bloch)
6. *Cirrhinus mngala* (Hamilton-Buchanan)
7. *Cirrhinus reba* (Hamilton-Buchanan)
8. *Cyprinus carpio* var *communis Linnaeus*
9. *Cyprinus carpio* var *specularis Linnaeus*
10. *Hypselobarbus curmuca* (Hamilton-Buchanan)
11. *Hypselobarbus dubius* (Day)
12. *Hypselobarbus kolus* (Sykes)
13. *Hypselobarbus pulchellus* (Day)
14. *Hypselobarbus micropogon* (Valenciennes)

15. *Hypselobarbus thomassi* (Day)
16. *Labeo bata* (Hamilton-Buchanan)
17. *Labeo calbasu* (Hamilton-Buchanan)
18. *Labeo fimbhatus* (Bloch)
19. *Labeo konitus* (Jerdeon)
20. *Labeo nigrescens* (Day)
21. *Labeo rohita* (Hamilton-Buchanan)
22. *Neolissochilus hexagonolepis* (McClelland)
23. *Osteobrama cotio cunma* (Day)
24. *Osteobrama vigorsii* (Sykes)
25. *Puntius arulies* (Jerdon)
26. *Puntius carnaticus* (Jerdon)
27. *Puntius dorsalis* (Jerdon)
28. *Puntius sarana sarana* (Hamilton-Buchanan)
29. *Schismatorhynchus nukta* (Sykes)
30. *Thynnichthys sandkhol* (Sykes)
31. *Tor Khudree* (Sykes)
32. *Hypophthalmichthys molitrix* (Valenciennes)
33, *Aorichthys aor* (Hamilton-Buchanan)
34. *Aorichthys seenghala* (Sykes)
35. *Mystus cavasius* (Hamilton-Buchanan)
36. *Mystus krishnensis Ramakrishnaiah*
37. *Mystus vittatus* (Bloch)
38. *Rita Kuturnee* (Sykes)
39. *Rita Pavimentaia* (Valenceennes)
40. *Ompok bimaculatus* (Bloch)
41. *Ompok pabo* (Hamilton-Buchanan)
42. *Wallage attu* (Schneider)
43. *Proeutropiichthys taakree taakree* (Sykes)
44. *Silonias children!* (Sykes)

45. *Pangasius pangasius* (Hamilton-Buchanan)
46. *Bagarius yarrellii Sykes*
47. *Carias batrachus* (Linnaeus)
48. *Heteropneustes fossilis* (Bloch)
49. *Glossogobius giuris* (Hamilton-Buchanan)
50. *Channa marulius* (Hamilton-Buchanan)
51. *Channa orientails Bloch & Schneideer*
52. *Channa striatus* (Bloch)
53. *Macrognathus* a/a/(Bloch & Schneidar)
54. *Mastacembelus* (Lacepede)

The west flowing rivers recorded 83 fish species belonging to 19 families and 7 orders, which includes the transplanted Gangetic and the exotic carps. *Hypselobarbus thomassi, Labeo nixcrsceiis* and *Glorias dussuntieri* have become "extinct" while 9 more fish components are under the "threat" of extinction. In the Bhadra reservoir 12 species of fishes including carp species have been recorded by Dr. Venkateswaralu.

Marine and Estuarine

The south-west monsoon from May/June until early September and the northeast from October until December predominates the climate of Karnataka and so influences the marine fisheries. The pelagic resources mainly Mackerals and Sardines are of dominating importance. Marine fish production in the state has shown wide fluctuations. The fishing, till the recent past, was confined to inshore areas. Efforts now are on to exploit deep sea resources to harvest Tuna, Pink Perch, deep Sea Prawns, etc. In general the fish catches are reduced presently compared with the production of 2.22 lakh tonnes during 1996-97 to 1.65 lakh tonnes during 1999-2000. The major fish catch consist of Mackerel, Sardines, Carangids, Seer, Scianids, Tuna, Ribbon fish, Anchovies, Soles, Flat-fishes, etc. There are 461 species of fishes, Soles, Flat fishes and Shrimps are recorded in the marine and estuarine areas.

On account of indiscriminate introduction of mechanized trawlers and purseseines, there is gradual declining trend in the fish catches. Though the State government has introduced "close season" for fishing during monsoon to protect the juveniles and brood fishes, there is no such restriction in the states of Goa and Kerala. Inspite of these restrictions imposed by the state, still, fishing with indigenous craft and gear in coastal waters continues. The fish population, therefore, is very much affected for not enforcing uniform conservation measures. Hence, there is dire necessity to enforce the legislation uniformly to the entire West Coast. The wetlands all along the coast are being converted into commercial farms for prawn farming which, to some extent, has affected the ecosystem. The list of economically important marine and estuarine fishes has been given in Table 23.2.

Table 23.2. Commercially Important Marine and Estuarine Fish of Karnataka

1. *Rastalliger kanagurtta*
2. *Sardinella longiceps*
3. *Sardinella fimbriata*
4. *Hilsa ilisha*
5. *Hilsa toli*
6. *Stromateus cinereus*
7. *Stromateus niger*
8. *Scomberomorous* spp.
9. *Euthynnus affinis*
10. *Tachysurus* spp.
11. *Lactarius lactarius*
12. *Trichiurus lopturus*
13. *Otolithes* spp.
14. *Protonibea diacanthus*
15. *Pseudoscianna diacanthus*
16. *Lizo macrolepis*
17. *Veelamugil scheeli*
18. *leiognathus* spp.

19. *Chironemus iysan*
20. *Chiroceentrus dorab*
21. *Engraulis* spp.
22. *Polyneemus* spp.
23. *Caranx* spp.
24. *Elocata niger*
25. *Sillago sihama*
26. *Chanos chanos*
27. *Etroplus suratensis*
28. *Pleuronectidae* spp.
29. *Pseudorhombus arsius*
30. *Cynoglossus* spp.
31. *Carcharhinus* spp.
32. *Sphyrna blochii*
33. *Pristis microdon*
34. *Trygon* spp.
35. *Sepia aculeata*
36. *Loligo duvauceli*
37. *Protunus* spp.
38. *Panuhrus* spp.
39. *Panaeus indicus*
40. *Metapeneeus dobsoni*
41. *Parapenaeopsis* spp.

In general, there is declining trend in the fish species, their population and growth rates compared to the sixties in all the river basins of the state. This may be attributed to various changes occurred in environmental parameters on account of certain developmental projects undertaken in the river valleys. Additionally, following problems encountered in the biotypes are also to be considered which too have contributed adversely to a large extent.

a. Indiscriminate Fishing

Some of the important pools and the natural fish sanctuaries like "falls", etc., where fish congregation takes place, are being dynamited very often during the lean seasons affecting the fish population in general and the brood stock too. This has been the case in all the major rivers of the state. In addition, indiscriminate fishing was resorted by the local and the migratory fishermen, especially in the reservoirs during the close-season, has also depleted the stock of fish population.

b. Introduction of Exotic Carp

Introduction of Common carp in certain man-made lakes during the fifties has left devastating impact on the indigenous fish species. In Krishnaraja Sagar reservoir common carp has dominated over all the other fish species indigenous to the Cauvery river and its reservoir. Same is the case in all the tanks of the state where in common carp has been introduced. Even the Gangetic major carp, as reflected in the catch composition of the inland waters of the state, is decreasing.

c. Stocking of Tilapia

Inadvertent introduction of Tilapia *Oreochromis mossambica* along with major carp seed has created a sort of destructive scene in all the tanks, rivers and the reservoirs of the state. The species has replaced many endemic fish species as also has thinned down their population. The size in the catches of Tilapia as observed in the recent past, are very much reduced.

d. Shrinkage of Resources mid Pollution

On account of encroachment of tank beds, the area of these is very much reduced as also, because of this process, number of tanks are lost for ever. Added to this, great number of tanks, especially in the malnad and semi-malnad region of the state are completely choked with aquatic weeds like water hyacinth, Typha, Impomea, Jussica, Cyperus, etc., which hampered the normal sustenance of fish life. Flow of pesticides from the agricultural fields to the water bodies and rivers too have affected greatly. Discharge of untreated effluents from industries and the domestic sewage into the tanks, rivers, estuaries and coastal waters have severely affected the fish fauna.

Approach to Biodiversity Conservation

Attention has been given to protect the bio-diversity of flora and land fauna to certain extent. But no conservation measures are undertaken for the sustainable conservation of bio-diversity of aquatic life. Therefore it is appropriate that suitable measures are taken to maintain the bio-diversity through *in situ* and *ex situ* conservation. This is due to the fact that about 38 fish species are under threat and 8 species have become extinct. Following measures are required to be considered to maintain the bio-diversity of finfish of Karnataka.

1. Despite legal prohibition, indiscriminate capture of spawners and juveniles and the use of dynamite and poisons have been going on unabated. The fishery regulations have played a subordinate role since they have been ineffective mostly due to enforcement problems. Adequate monitoring, control and surveillance are highly essential to sustain both the production and genetic diversity of the resources.
2. Monsoon being the breeding season for almost all the fish species, fishing should be prohibited to enable the stock to perpetuate their generation.
3. The leasing of river stretches for commercial exploitation of fish is in vogue. Since rivers form the cradle for breeding and recruitment which provide the feeding channel for natural stocking of indigenous fishseed in majority of the tanks, fish harvesting activity should be totally stopped during the prime season.

4. There are in the state, some declared fish sanctuaries in certain river systems, but the species and their number inhabiting these are markedly reduced on account of poaching, poisoning, dynamiting etc. Hence, there is urgent need to protect these sanctuaries as is being done in the case of wild-life sanctuaries.
5. Organise a fish faunistic survey of all the resources to assess the present "status" and to record the fish species coming under "rare, threatened, endangered" categories.
6. Systematic charting and ecological studies of deep pools which may be declared as sanctuaries for conservation of fish germplasm. Development of a strong database on fisheries and ecology of all major river systems is to be takenup. Improvement and provisions of spawning facilities from indigenous carp/cat fishes in areas where embankments for flood control have been constructed. Re-identification of breeding grounds and their protection through adequate legislation support.
7. Involve NGOs to take care of the river stretches for the protection of fish population and bio-diversity. Training the office bearers and NGO in management of fish congregation points in the riverine stretches during the lean season in regulating fishing and for providing feeding and protection throughout the year is needed.
8. To publish suitable literature for wide publicity for safeguarding the bio-diversity of aquatic life.
9. Close-season should be strictly enforced in the interiors of west coast by bringing uniform legislation to protect the marine bio-diversity.

10. There are several research institutes which are engaged in fisheries research, development, conservation and exploitation, i.e., Ecology and Environment Department, Central Fisheries Research Institutes, Marine Products Export Development Authority, Dept. of Irrigation, Universities, Department of Fisheries. These Institutions have their own programmes, but the problems addressed by the state is not taken into account. Therefore, there is a necessity of constituting a committee of all the concerned to evolve suitable strategies to protect the fisheries and their environment.

Reptiles and Amphibians

The Karnataka State is rich in herpetofauna. Although there is no correct figure that illustrates the number of species of reptiles and amphibians in the state, the two groups of vertebrates might together amount not less than 150 species. This is a rather conservative figure (Daniels, 2000). With the new addition of information on these groups from the researchers in the state, more and more species are getting added to the list of reptiles and amphibians. The topography, climate and the geographical spread of Karnataka has created a wide range of habitats that suit a rich assemblage of reptiles and amphibians. The extensive coastal habitats, the Western Ghats reaching heights of nearly 1800 in ASL, the many rivers, wetlands and the eastern plateau have enabled a tremendous diversity of natural and man-made habitats including beaches, estuaries, mangroves, rain forests, grasslands, deciduous forests, scrub and cultivated lands. Additionally, the more or less tropical climate has favoured the diversification of the herpetofauna. India is known to harbour 205 species of amphibians of which there might be more than 120 species of amphibians including caecilians, frogs and toads may be present in the state. Perhaps, this is 60% of the actual amphibian fauna known from the Western Ghats, thereby rendering the state richest in the region.

During the first half of the last century, professor C.R.N Rao of Bangalore had conducted extensive surveys of the amphibian fauna of Karnataka. He had collected and described a number of species during that time. Unfortunately, the specimens collected and deposited by Professor Rao have been lost. It has therefore not been possible for subsequent researchers to confirm the validity of at least 15 species described by him. Even if these are treated as 'uncertain', the amphibian fauna known from the state of Karnataka is far higher than any other comparable area in the Western Ghats.

The amphibian fauna of the state is interesting in many ways. The torrent toad (*Ansonia ornata*) an endemic species was first described from Bhagmandala in Coorg district about 100 years ago. Although currently the species is known from parts of Wyanad in Kerala, the largest population still exists in southwestern Karnataka. *NyctitHitrachits* is an endemic genus restricted to the Western Ghats. Of the 10 species of this hill stream frog described, eight are known from the state. Bush Frogs (*Philuatiis*) are amongst the most diverse genera of amphibians in India. Of these at least 23 species are known from the Western Ghats and many of them are endemic to Western Ghats. Fifteen species of bush frogs are known from Karnataka. Further, of the 17 species of Caecilians known from the Western Ghats, 3 genera and 14 species occur in the state. *Gegeneophis krislinai* is the most recently described species of Indian caecilians was based on a specimen obtained form Addoor in Dakshina Kannada.

New species of frogs have also been added from the state during the last two years. Amongst the new descriptions are *Microhyla sholigari* collected from B.R. hills and *Polypedates pseiidocruciger* collected from south-western Karnataka (Baindur). While there could be many more species of reptiles (not less than 80 species) in the state than there are amphibians, this group has not been equally well documented. We do know that there are 7 species of chelonians (turtles and tortoises) in Karnataka. Those are the olive ridley turtle (*Lepidochelys olimcca*), starred tortoise (*Geochelone elegans*), Travancore tortoise (*Indotestudo forsteni*), Leith's soft shell (*Aspideretes icithii*), flap shell (*L/ssrnn/s piinctata*), cane turtle (*Geomyda sylvatica*) and the pond turtle (*Melanochelys trijuga*). As with other parts of the Indian coasts, there might be stray records of sea turtles other than Olive Ridley which have been overlooked.

Human persecution has reduced the natural populations of the marsh crocodile (*Crocatylus palustris*) in the state. Nevertheless, there are still protected populations of the species in the river Cauvery and other major rivers in the north. Dr. Basavarajappa reported on breeding ground of freshwater crocodiles in the Kali River near Dandeli. The people from nearby villagers are involved in the poaching of crocodile eggs. Still to this date, the place is not declared as protected area. Few recent reports are available on the exact status of the crocodile in Karnataka.

The diversity of lizards in the state is less understood. The largest and widespread species of lizard in the state is the common monitor lizard (*Varamis Iviigalciisis*). Besides the monitor lizard, there are the Indian chameleon (*Chameleon zeylanicus*), 4-5 species of calotes, one species of flying lizard (*Draco dussumieri*), one species of fan throated Lizard (*Sitana ponliceriami*), two species of rock lizards (*Psamntophilns darsalis* and *P. blanfordanus*), geckoes in the genera *Hemidactylus, Cnemaspis, Cyrtodactylus, Hemiphyllodactylus* etc., and a number of skinks (*Mabuya, Riopa* and others). The exact number of species in each of these genera within the state is still unknown.

Undescribed species of lizards are being discovered by recent studies. For instance, a small gecko collected in the Indian Institute of science campus, Bangalore, a few years ago has been recently described as *Hemiphyllodactylus aurantiacus.*

Geckoes in this genus are generally forest dwellers known from parts of the Western Ghats, Eastern Ghats, North-eastern India and South-east Asia. Also, one of the most species rich genera of lizards viz., *Cnciuaspis,* commonly called dwarf geckoes, contain the maximum number of species endemic to south India. Very little is known of these small forest geckoes which are largely tree dwelling and diurnal. Dwarf geckoes are widespread in the hills and forests of Karnataka.

Snakes in the Karnataka are as little studied as the lizards are. What we know is that there are still small populations of the Indian rock python (*Python inolfinis*) and king Cobra (*Ophiophagus Hannah*) within the state. The former is widespread, whereas the King Cobra is restricted to the Western Ghats and Dakshina Kannada district. Venomous snakes, apart from the King Cobra, known from the state include the Indian cobra (*Najn naja*), common Krait (*Bungarus caeruleus*), Saw scaled viper (*Echis carinatus*), Russell's viper (*Vipcra msselli*) and Pit vipers in the genera *Trimeresurus* and *Hypnale.* Very little is known about the diversity of sea snakes along the coast.

Eryx zvltilakeri a species of sand boa was first described about 10 years ago after specimens were collected from Mangalore. In the same area, an unusual and unmarked colour variant of the Russell's viper has been collected. More careful investigations might reveal that this viper is a new species too. Interestingly, the common krait which is normally found in plains and cultivated areas, occurs in the evergreen forests in Uttara Kannada district also.

Amongst the arboreal snakes are *Ahaetualla nasutus, A. Pulverulentus, Chrysopelea aruata, Dendrelaphis tristis,* and *Boiga forsteni.* Uropeltidae, the unique and endemic family of burrowing snakes restricted to the humid tracts of south-western India and Sri Lanka, is also well represented in the state. It is however, cryptic and its diversity is least understood.

The above discussion was intended to merely highlight the diversity of the herpetofauna in the state of Karnataka. Very little is known of the ecology of the different species, their status and geographical range within the state. Habitat loss and direct killing by humans (especially larger reptiles) are continuously taking a heavy toll in the herpetofauna. During the rains, a number of species are killed on roads, while attempting to cross from one habitat patch to the other or from one side of the road to another side.

Research should focus on updating the species lists and adding information on geographical range, habitat requirement and status. Species under heavy human pressure need special attention. Herpetofaunal sanctuaries need to be considered. Potential localities in Karnataka are within the forests of Dakshina Kannada (eastern side) including Kudremukh, Agumbe, Neria Estates, Dharmasthala, Subramanya, Sringeri and the intervening hill ranges. This complex of undulating hills, grasslands and evergreen forests seem to be the best habitat for both amphibians and reptiles in the state.

Attention should also be given to educate the people in the state regarding the importance of these species in maintaining ecological balance. Reptiles and amphibians, in general are the most misunderstood of organisms. They are persecuted much more than any other group of animals due to the scores of myths and susperstitions that prevail amongst the rural public.

The Caecilians of Western Ghats

These are the legless amphibians quite characteristics of Western Ghats. Among 17 known species of India 14 are found in Western ghats and all are endemic. A field guide has been developed by Dr. Gopal Krishna (1998) of Shringeri for the identification of caecilians based on a survey made all over the Western Ghats (Figs. 23.0, 23.1 and 23.2).

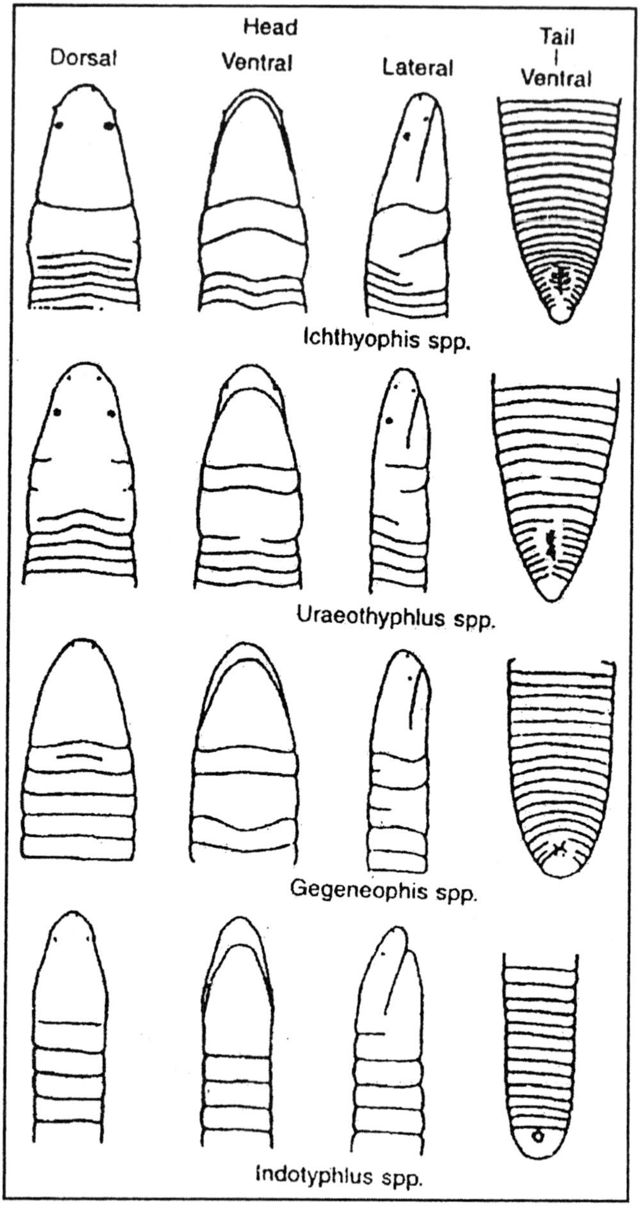

Fig. 23.1 : Distinctive features of four genera of caecilains showing relative position of nostrials, tentacles, eyes, nuchal groves and vent.

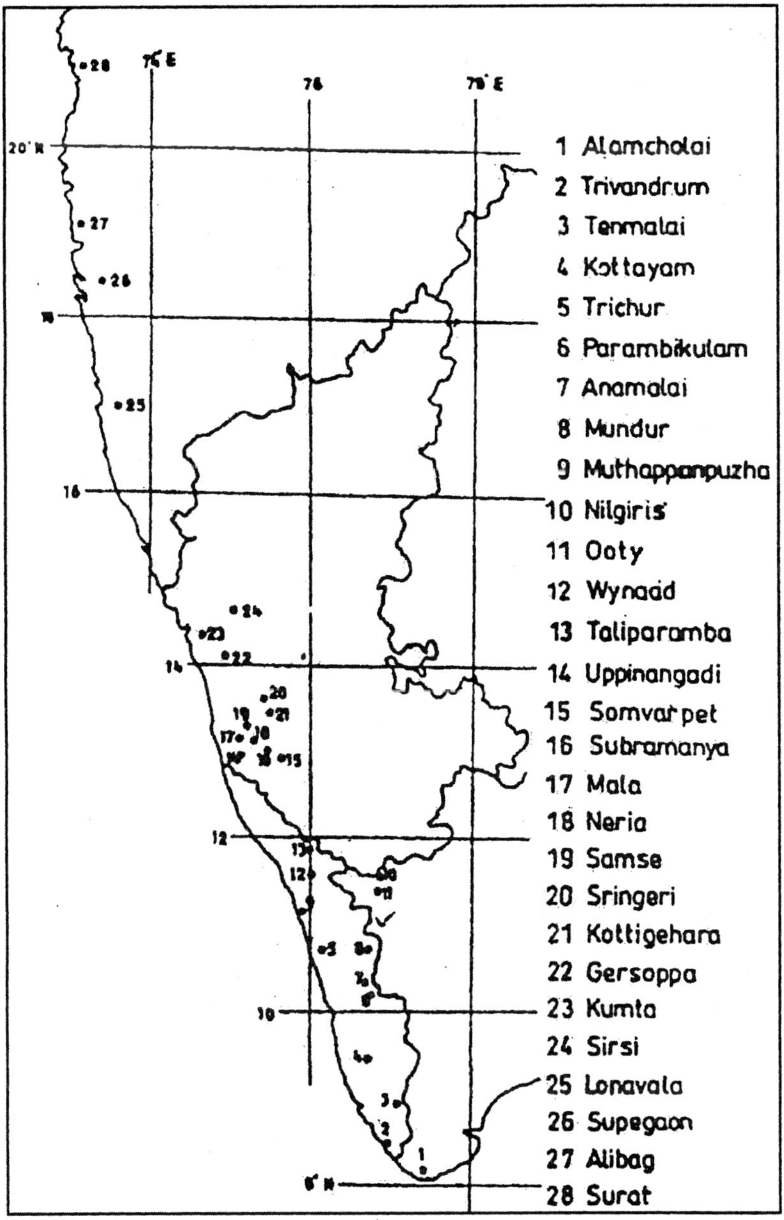

Fig. 23.2 : Distribution of Caecilains in India

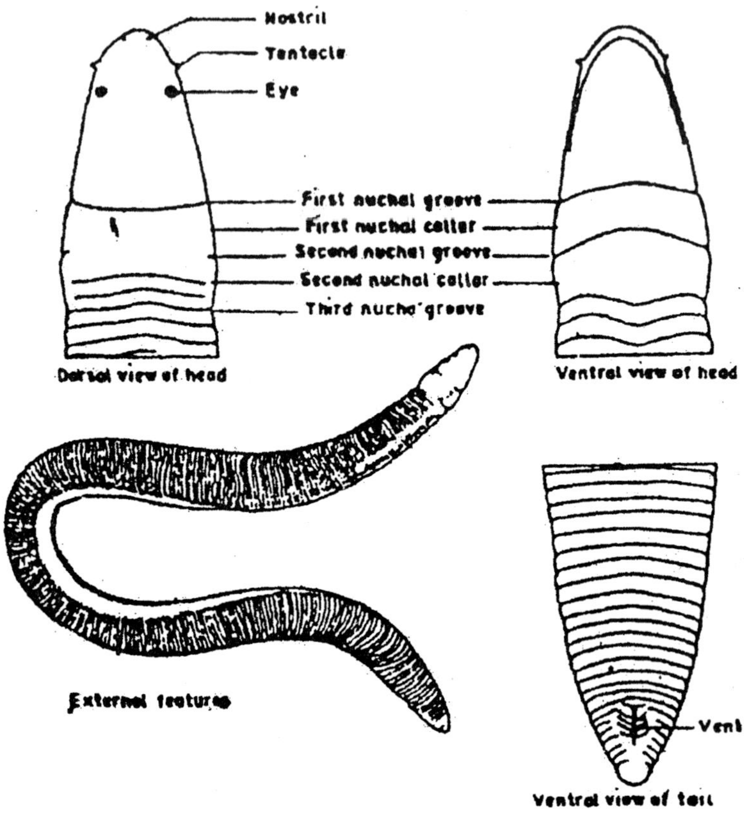

Fig 23.0. Caecilain Morphology Showing the Key Characters used in the Identification of Species

These amphibians due to burrowing, secretive and nocturnal habit, encounter in the field only occasionally. Due to this behaviour these popularly referred as quite rare species though these are quite common in specific habitats. The first authentic report is by Taylor (1968). He has published a monograph on the Caecilians of the world giving a summary of the existing literature and descriptions of many taxa.

India is known to harbour around 205 species of amphibians of which 120 species occur in the Western Ghats (Inger and Dutta, 1987). So far 17 species of caecilians are described from India (Table 23.5). Out of these 14 species are reported by Bhatta from the Western Ghats, which are endemic to the area.

Taxonomical studies on caecilians in India is limited (Gundappa *et al.*, 1981). Daniel is also published a field guide to the amphibians of Western Ghats of India, it mainly describes the anurans. The Indian Caecilians are classified into three genera (Table 23.3).

Table 23.3. Taxonomy of Indian Caecilians

Family	*Genus*	*Species*
Ichthyophiidae Taylor	*Ichthyophis* Fitzinger	*beddomei* Pater* *bombayensis* Taylor* *glutinosus* (Linnaeus) *longicephatus* sp. Nov* *malabarensis* Taylor* *peninsularis* Taylor' *sikkimensis* Taylor *subterrestris* Taylor'fryco/or Annandale*
Uraeotyphildae	*Uraeotyphlus* Peters'	*Malabaricus* (Beddome)*
Nussbaum		*menoni* Annandale* narayani Seshachar* *oxyurus* (Dumiril and Bibron)*
Caeciliidae Refinesque	*Gegeneophis* Peter*	*carnosus* (Beddome)* *fuller!* (Alkock) *ramaszvamii* Taylor* *battersbyi* Taylor
	Indotyphlus Taylor	

*Endemic to India; 'Endemic to Western Ghats.
Source: Bhatta GK, 1998.

Description of the animals

Since having rings on the bodies, these are generally mistaken with gaint earthworms or snakes. Caecilians; posses small eyes, sometimes covered with skin, a pair of nostrils, at the tip of the snout and a pair of sensory tentacles on either side of the head. The terminal or subterminal mouth has two rows of teeth on the upper jaw and one or two row on the lower jaw. The teeth are modified for holding and cutting the prey.

The tongue is fused with the floor of the buccal cavity and cannot be protruded. The body colour varies considerably across the species, having shades of violet, brown, grey and black. Some have bright yellow lateral stripes. The skin is smooth and slimy and in many there are small scales embedded in it. Skin secretes mucous which is toxic to predators. A short tail may or may not be present. The anal opening is longitudinal or transverse. Externally male and female look alike. Males have a single protrusible copulatory organ called phallodaeum, which is inserted into the vent of the female during copulation and hence internal fertilization occurs. All the caecilians so far reported from India are terrestrial and oviparous. The female lays a clutch of eggs in a burrow near a stream. The number of eggs also varies according to the species. The eggs are connected to each other with the help of gelatinous string. The mother coils her body around the eggs clutch until hatching. The larvae have external gills at the time of hatching and soon wriggle into the water where they loose the gills and metamorphose into terrestrial adult forms. Most of these Caecilians eat earthworms as their main diet and a few eat termites and other feed on soil invertebrates. These become food for snakes, birds, wildboars and also to domestic chickens.

Habitat

These are found at elevations ranging from sea level to 2500 m. AS1.. Generally inhabit loose soil rich in humus and are found in decaying wood. Moist leaf litter and in the piles of decaying coconut leaves at varying depths in the cavities at the bases of banana trees and aracanut gardens and coffee-tea plantations. They do occur in compost heaps, manure pits containing decaying leaves with dung. During wet season they come near to the surface of the soil and move freely in the night. The local folk kill them believing that they are snakes.

Table 23.4. Key for the Identification of Different Genera of Caecilians in India

	Ichthyophis	*Uraeotyphlus*	*Gegeneophis*	*Indotyphlus;*
Colour (in life)	Unicoloured forms are sober coloured with brown or greyish shade. Yellow lateral striped forms are violet brown on dorsal surface.	Olive violet, a steel grey on the dorsal and lavender on the ventral surface.	Greyish on the dorsal surface and light grey shaded on the ventral surface.	Pale flesh colored on both dorsal and ventral surfaces.
Mouth	Terminal	Recessed below the snout	Recessed below the snout	Recessed below the snout
Eyes	Visible	Visible	Not Visible	Visible through the skin
Tentacles	Conical, long, near edge of mouth, much closer to eye than to nostrils.	Flap like short, anterior to and below nostril, much closer to nostril than to eye.	Cone shaped, very short, close behind and below the nostril, nearer to tip of snout the to eye.	Cone shaped, short on a line between eye and nostril, closer to eye than to nostril.
Tail	Short and pointed	Short, tip whitish	No tail, body ends in a blunt shield.	No tail, body ends in a blunt shield.
Vent	Longitudinal	Longitudinal	Transverse	Transverse.

Table 23.5. Literature Available to the Distribution of Indian Caecilians

Species	*Localities of collection*	*References*
Ichthyophis beadomei	Kotigehar, Gersoppa, Kiggar,	Taylor 1960a, 1968
	Sringeri, Uppinangadi, Kumta,	Daniel 1963
	Sirsi, Mala, Neria-Karnataka	Tikader 1964
	Nilgiris-Tamil Nadu	Rahman and Rajagopral
	Munder-Kerala	1978 Balkrishna *et al.*,
	Supegaon-Maharashtra	1982a, b
		Bhatta 1986, 1968
Ichthyophis bombayensis	Waghii Surat-Bombay Province	Taylor 1960a, 1968
	Sirngeri-Karnataka	Daniel 1963
	Balkrishna *et al.*, 1982b	
	Revanasiddaiah *et al.*, 1982	
Ichthyophis longicephalus	Mundur, Muttappanpuzha-Kerala	Pillai, 1986
	Silent Valley-Kerala	Bhatta 1997
Ichthyophis malabarensis	Maduvangard, Maundur-Kerala	Taylor 1960a, 1968
	Mala, Sringeri, Sirsi, Subramanya,	Daniel 1963
	Kigga-Karnataka	Seshachar *et al.*, 1982
		Bhatta 1986, 1997
Ichthyophis peninsularis	Malabar, Vanjikadavu-Kerala	Taylor 1960a 1968
	Neria-Karnataka	Deniel 1963
	Alamcholai-Tamil Nadu	Jaisingh 1978
		Das and Whitaker 1990
		and author's collection
Ichthyophis sikkimentsis	Darjeeling-Sikkim	Taylor 1960a, 1968
Ichthyophis subterrestyis	Kottayam, Travancore-Cochin	Taylor 1960a
	Anamalai, Parambikulam-Kerala	Daneil 1963

	Alibag-Maharashtra	Taylor 1968
Ichthyophis tricolor	Maddathori, Peermade, Parambikulam	Taylor 1960a, 1968
	Travancore, Munder-Kerala	Daniel 1963
	Nilgiris-Tamil Nadu	Bhatta 1997 Taylor 1968
	Malabar-Kerala	
Uraeotyphlus malabaricus	*Ootacamund*-Tamil Nadu	
Uraeotyphlus menoni	Koduvalli, Cochin, Kottayam, Trichur-Kerala	Daniel 1963, Taylor 1968
Uraeotyphlus narayani	Kannam Ernakulam-Kerala	Daniel 1963
	Sringeri-Karnataka	Taylor 1968,
		Balakrishna *et al.*,1982a, b
		Bhatta 1986, 1997
Uraeotyphlus oxyurus	Taliparamba, Wynaad, Tinnivelly,	Taylor 1968
	Malabar, Allur, Cochin	
	Anamalai Hills-Kerala	
Gegeneophis camosus	Peria peak, Wynaad, Porunudi hills,	Seshachar1942
	Tenmalai, Trivandrum-Kerala Kotigehar,	Seshachar and Ramsawami
	Malige Range-Karnataka	1943
		Daniel 1963, Taylor 1968
		Revanasiddaiah *et al.*, 1982
Gegeneophis fulleri	Kuttal-Cachar	Taylor 1968
Gegeneophis ramaswamii	Tenmalai, Trivandrum,	Taylor 1964,1968
	Bonocord-Kerala	Bhutta, 1997
Indotyphlus battersbyi	Khandala, Lonavala	Taylor 1960b, 1968,1970
	Supegaon-Maharashtra	Bhatta, 1997

Source: Bhatta, O.K., 1998

There is wide spread confusion in identifying Caecilians; species. It is because of lack keys and field guides and lack of experience. It is also due to non-availability of holotypes paratypes in Indian museums for confirmation. General morphology of a Caecilian is shown in Fig 23.0. The distinctive features (Fig 23.1) and proper keys (Tables 23.4 and 23.5) for identification of Caecilians is made available.

Identification of Caecilian Genera

It is ideal to search for caecilians in the rainy months, after ram percolates considerably so as to originate subsoil springs. These are generally nocturnal. During day time generally they shelter beneath the decaying vegetation or similar ecological situations. After siting an individual one has to first ascertain that the animal is definitely a Caecilian. by confirming that it possess the rings (annuli) on the body. Taking the animal in the hand it is necessary to identify the species to which it belongs. They wriggle out to escape from our hand but do not bite. The species belonging to genera *Ichthyophis* and *Uraeotyphhis* possess pointed tail and longitudinal vent (Refer Fig 23.1) where as those of the genera *Gegmeophis* and *Indotyphlus* have blunt terminus with transverse vent. With the exception of unicoloured forms of the genus *Ichthyophis* all other known species of Caecilians are not longer than 30 cm in body length.

Caecilians of Western Ghats

Ichythyophis glutinosis is absent in India however it is wrongly reported as present in our country. The morphological details of Indian Caecilians are described in terms of their body colour, yellow banding pattern, relative position of eye, nostril, tentacle and the nuchal groove.

1. Ichthyophis beddomi

The body is violet brown on the dorsal side and light brown on the ventral side. A yellow lateral stripe passes along the length of body from mouth to tip of the tail. The stripe bifurcates near the mouth and one branch reaches the tip of the snout along upper lip and the other reaches the tip of the chin along the lower jaw. The yellow strip widens just opposite to first collar, then becomes narrow and again widens and continues till the tip of the tail. The body is short and broad, eyes are distinct. Tentacular aperture is very close to lip. Its distance from the eye and nostril is almost equal. Nostrils nearly terminal, but barely visible when viewed from above the head, the snout extends slightly beyond the mouth. The first nuchal groove is distinct on both dorsal and ventral sides. Second nuchal groove is not clear on dorsal surface and the third nuchal groove is distinguishable on the dorsal side but not on the median ventral region. The second collar groove bears two well defined folds dorsally. It occurs in Karnataka, Kerala and Maharashtra state.

2. *Ichthyophis bombayensis*

It is large unicoloured species with elongate tapering tail. The body is dark brown above and somewhat lighter brown below. Lateral yellow bands are absent. The head is small and narrower than the body. The eyes are distinct and surrounded by a whitish ring. Tentacular aperture is close to the lip. It is nearest to the eye than the nostril. The first nuchal groove is distinct ventrally at the sides and preceded dorsally by a strongly curved fold on the back of the head. Second nuchal groove is clear ventrally and laterally but not dorsally. Third groove crosses on the dorsal side visible laterally and partly crosses the venter. It occurs in Gujarat and Karnataka.

3. *Ichthyophis longicephalus*

The body is dark violet brown on the dorsum and somewhat lighter on the venter. A yellow lateral stripe extends from tip of the tail to posterior margin of second collar. Small yellow patches occur at the sides of the second collar, first collar and on the angle of jaws. On the venter is annulus broken in the middle and results in the longitudinal midventral line extending from collar to vent. Body is short and broad. Head is comparatively longer. Eyes are distinct, tentacular aperture lies close to lip. It is much closer to the eyes than to the nostril. Nostrils are nearly terminal and visible from the above. The snout extends a little beyond the mouth. The first nuchal groove is clear ventrally and laterally and dim dorsally. The first collar is completely fused with the second on the dorsum. The second nuchal groove is distinct on the ventral surface and extends up laterally upto the level of angle of mouth. The second collar bears two incomplete folds dorsally. It occurs in Kerala.

4. Ichthyophis malabarensis

This is the largest known Indian Caecilian species. It is unicoloured, dark brown above and light cream white coloured below. The tip of the tail is dark. and nearly black from vent to tip. No lateral yellow strip. Head is narrower than the body. Eyes are visible. Tentacular aperture near the lip and is closer to eye than to nostrils. The first nuchal groove is clear ventrally and becomes slightly angulate dorsally. The second nuchal groove is convex below and distinct on the sides of the neck. The third nuchal groove is clear on the sides of the neck and dim on the dorsal surface. There are three transverse folds on the dorsum of the second collar. It occurs in Kerala and Karnataka.

5. Ichthyophis peninsularis

Unicoloured form, dorsum is grayish lavender and the ventral side is cream coloured. No lateral yellow stripe. It is a large species with broad and short head and relatively long tail. Eyes are visible, tentacles lie near to eyes than to nostrils. The first nuchal groove is clear dorsally, laterally and ventrally. The second nuchal groove is well defined on the ventral surface and on the sides of the head. The third nuchal groove is more or less distinct, except medially below. It occurs in Kerala, Karnataka and Tamil Nadu.

6. Ichthyophis subterrestris

A moderate sized species with dark violet lavender colour on the dorsum and lavender brown on the ventral sides. There is no lateral yellow stripe. The head is elevated posteriorly and eyes are distinct. The tentacle is near to the eye and close to the lip. The first nuchal groove is distinct on both dorsum and the venter. The second groove is not clear on the dorsum and it is found in Kerala.

7. Ichthyophis tricolor

Body is violet brown above, a yellow lateral stripe extends from the head to the tip of the tail not broken on the neck but widens on the collars. At the angle of the jaws it forks and the yellow band continues along the jaw up to the eye on the dorsal surface on the snout. A broad white ventral stripe is present on the venter. Head is with slightly projected snout. Eyes are distinct. Tentacle at the edge of the lip, near to the eye than to the nostril. The first nuchal groove is distinct on the venter passes at the sides, but doesn't meet on the mid dorsal line. The second nuchal groove is clear on the ventral side and at the sides of the neck but not dorsally. There are two transverse folds on the dorsum on the second collar. It is found in the Kerala state.

8. Uraeotyphlus malabaricus

Body is short and stout, violet coloured on the dorsum and lighter shade of violet on the venter. Eyes are distinct and surrounded by a cream ring. Tentacles sparsely visible from above, near to the nostril than to the eyes. It is closer to the mouth than to nostril. Small cream eyes are present at nostril, tentacle, tip of the snout and its underside. The upper and lower lip and the two jaws are also cream in colour. A cream white spot is present at the vent. The tip of the tail is whitish. The first collar is very dim not seen from above, and vaguely evident from ventrally. The first collar is fused with second collar. Second collar is more clear, incompletely grooved posteriorly with slightly visible transverse fold across the dorsum and is found in Kerala.

9. Uraeotyphlus menoni

Medium sized species, slate grey on the dorsum and whitish blotched with grey on the venter. Eyes are distinct and surrounded by a narrow white ring. Tentacle not visible from above, closer to the nostril than to the eye. Nostrils are distinctly visible from above. The head is light violet coloured, tip of the snout above and below is cream white with light mottling on top of the head. The chin and throat are with indefinite greyish marks. The top of the tail is cream coloured.

The first nuchal groove is less distinct above and clearly defined below. The second is not clear dorsally but visible laterally and ventrally. The second collar is well defined dorsally and laterally but ventrally, it is fused with the first body annulus and is found in Kerala

10. Uraeotyphlus narayani

Steel grey above, pale and flesh coloured on the venter. In between chin and tail a median greenish line is present. Eyes are distinct with a whitish aerola. Nostrils are visible directly above the head. The distance from the eye to nostril and eye to tentacle is almost same. Tentacular aperture is below the nostril and not visible from above directly. Tip of the tail is whitish. The tip and ventral part of the snout is cream. No light spot at tentacle or nostril. The first nuchal groove is distinct ventrally and laterally and less clear dorsally. The second nuchal groove clear ventrally and extends up on the sides to the level of mouth. The third nuchal groove is visible on the dorsum and sides of neck and is found in Kerala and Karnataka.

11. Uraeotyphlus oxyurus

A thick bodied species with narrower head and a short tail. Body is moderately dark brown. Eyes are very small, but distinct with a light white ring. Tentacle below nostril, closer to nostril than to eye. Nostrils are visible from directly above head. The tip of the tail is white. The chin and the throat are very light brown. The first nuchal groove is very dim across the head. The second collar is fused to first primary fold on the venter. A single fold lies across the back part of the second collar and it is found in Kerala.

12. Gegeneophis camosus

A small species with body flesh coloured on both sides, terminal width less than body width. Body ends in blunt shield. Eyes are invisible. The tentacular aperture is below and behind the nostril. Nostrils are nearly terminal and not visible directly above head. The terminus is narrowed. The first nuchal groove is clear on the venter and dim on the dorsum. The second collar is distinct above, partly divided by dorsal transverse groove, while on the ventral side it is partly fused with primary annulus. It is found in Kerala and Karnataka

13. *Gegeneophis ramaswamii*

Lager species with greyish on the dorsal side and light grey on the ventral side. Terminus of the body is wider than elsewhere, which ends bluntly. Tentacles lie behind and below the nostril, not visible from directly above the head. Eyes are invisible. The first and second nuchal groove are distinct on the dorsum, venter and at the sides. There is a short transverse groove on its first collar. It is found in Kerala.

14. *Indotyphlus battersbyi*

It is a caecilian with slender elongate body, pale flesh coloured on both above and below. Eyes are concealed and feebly visible through the skin. Body ends in blunt shield. The first nuchal groove is distinct on the dorsum, while the second is clear on both above and below. It is found in Maharastra. The Caecilians are endemic to moist, swampy and high attitudinal areas of Western Ghats. These are not reported from the planes of Karnataka.

Avifauna

Unlike many States of India, Karnataka still has vast tracks of natural forest and there are some places where still no ornithologist has visited. The state embodies a diverse types of country sides and forests teeming with a mosaic of habitats, providing a heaven for many rare and endemic birds. The Western Ghat is an important habitat to a large number of forest loving bird species. But, such richness of natural heritage cannot be taken for granted. Without good management and implementation of environmental and wildlife laws, invaluable resources that are existing today may cease to do so. Located in the west-central part of peninsular India, Karnataka has 31 forest types. The diverse forest types include the Hill top tropical evergreen forests, the West coast tropical evergreen forest, Secondary moist bamboo brakes and the Secondary dry deciduous forests (Table 23.6). Five National Parks and nineteen Sanctuaries have been notified under the Wildlife Protection Act of 1972, in Karnataka between 1972 and 1989, which harbor a large variety of bird fauna (Table 23.7). Twenty Seven Species of Endemic birds of Peninsular India have been reported in the state (Table 23.8). The total number of birds found in Karnataka is about 500.

The Department of Forests, Karnataka State published a prelude to bird watching in Karnataka. It includes a list of 52 common birds of Karnataka. The list is as follows:

1. Crimson breasted Barbet
2. Black headed Oriole
3. White eye
4. Jungle Babbler
5. Spotted Babbler
6. Indian Robin
7. Magpie Robin
8. Indian Swallow
9. Pied Bush chat
10. Black headed Munia
11. Red Munia
12. White Backed Munia
13. Indian Myna
14. Grey Headed Myna
15. Hill Myna
16. Lotens Sunbird
17. Small Sunbird
18. Black Drango
19. Greater racket tailed Drango
20. White Bellied Drango
21. Common Wood Shrike
22. Black headed Cuckoo Shrike
23. Baybacked Shrike
24. Yellow Throated Sparrow

25. Scarlet Minivet
26. Small Minivet
27. Yellow legged green Pigeon
28. Ring Dove
29. Little Brown Dove
30. Sported Dove
31. Tailor Bird
32. Lesser Golden backed Woodpecker
33. Larger Golden backed Woodpecker
34. Pigmy Woodpecker
35. Yellow fronted Pied Woodpecker
36. Little scalybellied Woodpecker
37. Rufous Woodpecker
38. Black naped Flycatcher
39. Paradise Flycatcher
40. White browed fantail Flycatcher
41. White spotted fantail Flycatcher
42. Brown Flycatcher
43. Tickells blue Flycatcher
44. Chestnut headed Flycatcher
45. Blue bearded Bee eater
46. Bronge winged Jacana
47. Spoon bill
48. Dab chick or Little grebe
49. White backed Vulture
50. Black Winged Kite
51. White eyed Buzzard Eagle

Wetland Birds

Wetlands are among the most dynamic and complex ecosystems in the world and 66 per cent of continental extinction are due to aquatic taxa. This highlights the exceptional sensitivity of the wetland and freshwater ecosystems. 338 wetlands were inventoried as a part of the annual midwinter waterfowl census since 1987 in Karnataka (Table 23.9). There is a need to understand effectively and manage these wetlands in such a way that their bio-diversity is protected. The high impact threats to the bird diversity of Sanctuaries and National Parks (Table 23.7) and non-regulated wetlands are listed below:

1. Illegal occupation.
2. Unsustainable exploitation of forest resources.
3. Industries and labour camps.
4. Construction water related and aquaculture projects.
5. Unchecked deforestation.
6. Grazing by livestock.
7. Mining activities.
8. Forest fires.
9. Water and pesticide pollution.
10. In regulated hunting.
11. Poaching and filling of wetlands.

Table 23.6. 31 Forest Types Found In Karnataka (As per Champion & Seth's Classification)

1A/C3	Southern hill-top tropical evergreen forest.
1A/C4	West coast tropical evergreen forest.
1/E1	Cane brakes.
1/E2	Wet bamboo brakes.
1/251	Pioneer Euphorbiaceous scrub.
1A/C2	West coast semi-evergreen forest.
2A/2S1	West coast secondary evergreen Dipterocarp forest.
2/E3	Moist bamboo brakes.
2/E4	Lateritic semi-evergreen forest.

2/2S1	Secondary moist-bamboo brakes.
3B/C1	Slightly moist and moist teak forests.
3B/C2	Southern moist mixed deciduous forests.
3B/2 S1	Southern secondary moist mixed deciduous forest.
4C/FS2	Submontane hill valley swamp forest.
4E/RS1	Riparian fringing forest.
5A/Cl (b)	Dry teak forest.
5A/G3	Southern dry mixed deciduous forest.
5/DS1	Dry deciduous scrub.
5/DS2	Dry savannah forest.
5/DS3	Euphorbia scrub.
5/DS4	Dry grassland.
5/E4	Hardiwaickia forest.
5/E7	Laterite thorn forest.
5/E9	Dry bamboo brake.
5/1S1	Southern dry tropical riverian forest.
5/2S1	Secondary dry deciduous forest.
6A/Cl	Southern thorn forest.
6A/DS1	Southern thorn scrub.
6A/DS2	Southern Euphorbia scrub.
6/DS1	Cassia auriculata scrub.
8A/C2	Western sub-tropical hill forest.

During past two decades it is observed that in Karnataka state the population of common birds like House sparrow and common crow is declining. Once these were most common in urban and rural Karnataka. But now egrets and herons are most common. Why this sudden change is yet to be researched. Perhaps the reason for the decline of sparrow and crow may be the habitat loss and increase of concrete buildings even in village areas, which made the sparrow to loose its nesting grounds.

Table 23.7. National Parks and Sanctuary of Karnataka/Key Bird Species

National Park Sanctuary	*Forest types*	*Bird Diversity*	*Key Bird Species*
1	2	3	4
Anshi (250 Sq. kms)	Ever Green, Semi Ever green.	200-250	Malabar Pied Hornbill, Peafowl, Grey Jungle fowl.
Bandipur (874 sq. kms)	Dry Deciduous Scrub, Dry Teak Bearing Forest, Moist Teak Bearing Forest and Semi Evergreen Forest.	339	Peacock, Woodpeckers, Nuthatches, Flycatchers.
Bannerghatta (104 sq. kms)	Dry mixed deciduous forest, thorn forest, Moist mixed deciduous forest.	175-200	Peacock, White-browed Fantall Flycatchers White bellied tree pie.
Kudremukh (600 sq. kms)	Hill-top tropical evergreen, West Coast semi-evergreen, Shola.	250-300	Malabar Trogon, Great black woodpecker, Malabar pied hornbill, Orange breasted green Pigeon, osprey.
Rajiv Gandhi (643 sq. kms)	Tropical moist mixed deciduous forest, tropical Westcoast evergreen, scrub	275-280	Adjutant stork. Honey buzzard. Pelican, Woodpeckers (10 species).

Adichunchunagiri Peacock (84 sq kms)	Tropical dry deciduous thorn scrub; Eucalyptus Plantation.	50-100	Peacock, Hoope, Ring dove. Black Winged kite.
Arabithittu (13 sq. kms)	Scrub land, Eucalyptus, Sandalwood.	100-150	Peacock, Drongo, Grey jungle fowl.
Bhadra (492 sq. kms)	Dry mixed deciduous, Moist mixed deciduous, western sub-tropical hill forest, Teak plantation.	300-325	Avocet, Vulture, Malabar trogon, Great black woodpecker, Osprey.
B. R. Hills (539 sq. kms)	Moist teak bearing forest, dry deciduous scrub, Sub-tropical Hill forest, Silver Oak, Teak.	250-300	Fishing eagle, Large Indian parakeet, Wood owl, Trogon Velvet-fronted nuthatch.
Brahmagiri (181 sq. kms)	West coast tropical evergreen, semi-evergreen forest.	280-310	Coorg laughing thrush, Velvet fronted nuthatch, Fairy Blue Bird.
Cauvery (510 sq. kms)	Glossy land, Open cultivation, closed forest.	250-300	Nilgiri flycatcher, Serpent eagle, Black bellied teru.
Dandeli (834 sq. kms.)	Moist mixed deciduous, West-coast semi-evergreen forests, fruit-yielding trees, Teak, Bamboo, Eucalyptus.	250-275	Black woodpecker, yellow fronted pied, Woodpecker Hornbill, Nuthatch, Malabar Whistling thrush.
Ghataprabha (29 sq. kms.)	Swamp islands, Elephant grass, Bamboo, Acacia Plantations,	200-250	Demoiselle crane, Black bellied teru, Spoon bill.
Gudavi (74 sq. kms)	Moist mixed deciduous forest, Acacia and Bamboo plantations.	100-125	Jacana, Whistling teal, Ibis, Darter, Spoon Bill, Egret.
Melkote (49 sq. kms)	Dry deciduous scrub forest.	100-125	Peacock, Green Imperial Pigeon.

Mookambika (247 sq kms)	Hilltop tropical evergreen, West-coast tropical evergreen, Lateritic semi evergreen, West-coast semi evergreen, Secondary moist mixed deciduous Dry grassland.	280-300	Darter, Adjutant stork Comb duck, Malabar Pied Hornbil, Whistling teals, Vultures, Woodpecker, Flycatchers.
Nugu (30 sq. kms)	Dry mixed decidous, Dry deciduous scrub, Bamboo, Eucalyptus, Acacia plantation.	150-200	Fishing eagle, Yellownaped woodpecker Red spur-fowl lorikeet.
Pushpagiri (102 sq. kms)	Tropical evergreen, tropical semi evergreen, grassland, Most evegreen type habitats.	250-260	Hornbills, Trogon, Raptors, Nuthatch, Woodpeckers, Fairy blue Bird, Flycatchers.
Ranibennur (119 sq. kms)	Open Scrub Jungle Sandal, Eucalyptus, Albezia, plantations.	125-150	Great Indian bustard, Common sand grouse, White eyed buzzard, Black-backed woodpecker, White backed vulture, Black winged kite.
Ranganthittu (67 sq. kms)	Island, Dry deciduous scrub, dry tropical riparian, Tropical riparian fringing forests, Thorny bamboo, Pandanus, Pongamia.	100-150	Darter, Sponbill, Spot bill duck, Painted-stork, Shag, Osprey, Stone plover, Indian river tern, Whistling teal, Herons, White ibis.
Sharavathi (431 sq. kms)	Lateritic semi-evergreen, Hilltop tropical evergreen, small stretches of west coast tropical evergreen,	300-325	Adjutant stork, Darter, Malabar pied hornbill, Lorikeet, Whistling teal, Yellow throated sparrow,

	west coast semi-evergreen forests.		Great Black woodpecker.
Shettihally (395 sq. kms)	Moist mixed decidious, West coast semi-evergreen, Tropical dry deciduous, Silver oak, Eucalyptus plantations.	200-250	Adjutant stork, Darter, Malabar pied hornbill, Woodpecker.
Someshwara (88 sq. kms)	West coast tropical evergreen, West coast semi-evergreen, Secon moist mixed deciduous, Dry grassland, Cashew, Teak, Eucalyptus plantations.	260-270	Darter, White-cheeked, bulbul, Flycatcher, Malabar pied hornbill, Lorikeet, Whistling teal, Yellow fronted pied woodpecker.
Talakaveri (105 sq. kms)	Tropical wet evergreen forest, tropical semi-evergreen, grassland.	250-275	Flycatcher, Malabar Trogon, Hornbills, Raptors, Nuthatch, Woodpeckers, Fairy blue bird.

Another interesting study conducted by the author, is that out of the four species of true baya birds found in India, only three were reported from Karnataka. The two species viz., *Ploceus Philippines* and *Ploceus bengalensis* are commonly found whereas Ploceus manyar is rarely cited. There is need to consider this species as one of the most endangered baya bird in Karnataka based on the survey in other States. Various factors threatening the existence of wetlands in Karnataka are shown in Table 23.9.

Table 23.8. Endemic Birds of Peninsular India

SI. No.	*Name*	*Distribution*
1.	Blue Winged parakeet	W.G.
2.	Small Green-billed Malkoha	P.I. & S.L.
3.	Sirkeer Cuckoo	P.I. & S.L.
4.	Wnite-rumped Spine-tailed Swift	N.E. & P.I.
5.	Malabar Pied Hornbill	W.G. & N.E.
6.	Small Green Barbet	P.I.
7.	Crimson Throated Barbet	W.G.
8.	Malabar Crested Lark	W.I. & W. G.
9.	While-bellied Tree Pie	W.G.
10.	Grey-headed Bulbul	W.G.
11.	White-browed Bulbul	P.I.
12.	Yellow-throated Bulbul	P.I.
13.	Black-headed Babbler	W.G.
14.	Rufous Babbler	W.G.
15.	White-headed Babbler	P.I.
16.	Laughing Thrush	W.G.
17.	Nilgiri Laughing Thrush	Nilgiri
18.	Jerdons White Breasted Thrush	Coorg
19.	Black & Orange Flycatcher	W.G.
20.	White-bellied Blue Flycatcher	W.G.

21.	Nilgiri Verditer Flycatcher	W.G.
22.	White-spotted Fantail Flycatcher	P.I.
23.	Broad Tailed Warbler	W.G.
24.	Rufous Bellied Shortwing	W.G.
25.	Nilgiri Pipit	W.G.
26.	Loten's Sunbird	P.I.
27.	Rufous Bellied Munia	W.G. & E.G.

W.I. = Western India, N.E. = North East, W.G. = Western Ghats. P.I. = Penninsular India, S.L. = Sri Lanka

Table 23.9. Threats to Wetlands Covered in Karnataka

Sl. No.	*Threats and Uses*	*No. of Wetlands*	*%*
1.	Free from Threats	10	(03%)
2.	Sedimentation	146	(44%)
3.	Excessive Growth of Vegetation	72	(22%)
4.	Cutting/Clearing of Vegetation	62	(19%)
5.	Eutrophication	22	(07%)
6.	Agriculture Along Drying Margins	145	(43%)
7.	Excessive Cattle Grazing	33	(10%)
8.	Pollution by Domestic Sewage	51	(15%)
9.	Solid Waste Dumping	8	(03%)
10.	Industrial Wastes	14	(04%)
11.	Pesticides & Fertilizers	17	(05%)
12.	Mudlifting & Brick Making	78	(23%)
13.	Hunting & Tapping Birds	67	(20%)
14.	Fishing	130	(39%)
15.	Partial Reclamation (includes 2 Fully Reclamined Wetlands)	82	(25%)
16.	Dam Construction	1	(.3%)
17.	Recreation & Tourism	6	(02%)

Mammalian Diversity In The Protected Area Network Of Karnataka State

Mammalian populations in the wild are now restricted to the remote forests of Dandeli wildlife Sanctuary in Uttara Kannada district in north. From these and adjoining forests mammals moved freely, until 1912 into the forests of erstwhile near Mysore. They are now trapped in isolated pockets, due to loss of forests. There are also the much fragmented populations in Bhadra Wildlife Sanctuary (now under Project Tiger and elephant schemes, since 1998), Kudremukh reserve, Agumbe and Kardibetta and Kodagu (formerly Coorg) forests. Ryley (1913) records mammals, as being periodically numerous in north and east Coorg, moving freely back and forth from Mysore and the Wayanad forests.

The very important mammal populations range more or less in a contiguous belt of forests from Nagarahole reserve and across the Kabini reservoir, down south that lies Bandipur reserve, far east lies Biligirirangana hills Sanctuary, that connects Mahadeswara hills, all these together forming a large expanse of protected areas, extending more than 2,500 km. This largest human modified stretch of "biological diversity' forms a part of "Nilgiri Biosphere Reserve". Here some of the mammalian populations, such as elephants and gaurs, overlap each other. Animals from adjoining forests, of Wayanad wildlife Sanctuary of Kerala and Madumalai Wildlife Sanctuary and National Park of Tamilnadu also keep shuffling freely during their seasonal movements. Other larger mammals and lesser known mammals that remain as resident populations in their respective protected areas are responsible in maintaining a in situ healthy bio-diversity.

In plains to the east of Western Ghats there are few isolated wildlife sanctuaries with isolated mammalian populations, such as the blackbuck, the gazelles and the wolves. It is said that the famous Indian Hunting Cheetah, once roamed plains and grasslands of Raibag and Ranibennur is extinct some 200 years ago. There are many mammals that face immediate threat of extinction.

Table 23.10. Avifauna recorded from Western Ghats of Karnataka (Annon, 1995; AH, 1996; Gina, 2000)

R = widespread resident, r = very local resident; W widespread winter visitor; w resident; ? = status uncertain = sparse winter visitor, P = widespread migrant; p = sparse migrant; V = vagrant, $_{rt}$ = n need

No.	Order	Family	Species	English Name	Local name	Local Status	Global Status
1.	Ansenformes	Anatidae	*Anas* crecra	CommonTeal	Solan, Hakki (k)	W	
2.			Neffapus *coromandeliamis*	Cotton Teal	Girri, Girria, Ghji (h)	R	
3.		Dendrocygnidae	*Dendrxygna javanica*	Lesser Whistling Teal	Seelhi Seek kahi (h)	R	
4.	Apodtformes	Apodidae	*Apus affinis*	Little (House) Swift	Ababeel, Babeela (n)	R	
5.	Bucerotiformes	Bucerotidae	*Anthracocems coronatus*	Matabar Pied hornbill,	Dhan chirl (h)	R Rare?	LR/nt (IUCN)
6.	Ciconiitormes	Accipitndae	*Accipiter* badius	Shikra	=shikra; = chipka (h)	R	
8.			*Aquila vindhiana (rapax)*	Tawny Eagle	Ukaab, Ragar (h)	R	
9.		*Butastur leesa*	White-eyed Buzzard	Tisa (h)		R	
10.			Circaetus galliucs	Short-toed (Snake) Eagle	Malpatar (k); Saampar (h)	R	
11.			*Circus macrourus*	PaleHamer	Dastamal, pattai (h)	W	LR/ntq(IUCN)
12.			*Gyps bengatersis*	White-backed (Bengal) Venture	Gidh (h)	R	LR/ntq/UCN)
13.			*HaCstur Indus*	Brahrminy Kite	Garuda (k); (h) Brahmani keel	R	
14.			*Milvus mtgrans*	Black/Common Pariah Kite	Cheel (h)	R	
15.			*Neophron percmpterus*	Egyptaki/White Scavenger Vulture	Safed gkth, Kal kurgh (h)	R	
16.			*Pandion haliaetvs*	Osorey	Machhariya, Machlimar	rW	

17.			*Sarcogyps calvas*	Black/King Vulture	Raj/Mulla gidh. Bhaonra (h)	R	LR/ntJUCN;
18.			*Spilomis cheela*	Crested Serpent-Eagle	Furj baz, Dogra cheei (h)	R	
19.			*Spizaetvs cirrhatus*	Crested Hawk-Eagle	Shah baaz (h)	R	
20.		Anhingidae	*Anhinga rufa*	(Oriental) Darter ot Snake Bird	Penwa Pan dubi) Banbe (h)	R	
21.		Ardeidae	*Ardea aft/us*	Large (Great) Egret	Bada. Bagla (h)	Rw	
22.			*Ardecla grayii*	Pond-Heron or Paddy Bird	Bagla (h)	R	
23.			*Bubulcus ibis*	Cattle Egret	Surthia bagla (h)	R	
24.			*EgreKa ganetta*	little Egret	Karchia bagla. Kitehia (h)	R	
25.			*Nycticorax nycticorax*	(Black-crowned) Night-Heron	Waak, Kwaak, Komi (h)	R	
26.		Ciconiidae	*Anastomus oscitans*	Openbill Stork	Gungla, Ghonghila, (h)	R	LR/nl (IUCN)
27.			*Ciconia episcopus*	(Woolly) White-necked Stork	Laglag (h)	R	
28.			*Ephippiorhynchus asiaticus*	Black-necked Stork	Loharjang Banaras(h)	R	
29.	(Charadrifformes)	Charadnidae	*Charadrius dubius*	Little Ringed Plover	Zierrea, Merwa (h)	RW	
30.			*Himantopus himantopus*	Black-winged Stilt	Gaz paon, Tinghur (h)	R	
31.			*Vanellus indicus*	Red-wattled Lapwing	Titeeri, Titai Titi Titor	R	
32.			*Vanellus malabaricus*	Yellow-wattled Lapwing	Zirdi (h)	R	
33.		Fateonidae	*Faleos biarmicus jugger*	Laggar Falcon	Laggar, Jaggar (h)	R	
34.		Jacanidae	*Hydmphasianus chirurgus*	Pheasant-tailed Jacana	Piho, Pihuya (h)	R	
35.			*Metopidius indicus*	Bronze-winged Jacana	Dal pipe, Karatjya	R	
36.	(Charadriiformes)	Laridae	*Sterna aurantia*	River Tern		R	
37.		Phalacrocoracidae	*Phalacmcorax carbo*	Large (Great)CormoranT	Pan kowwa (h)	RW	
38.			*Phalacrocorax niger*	Little Cormorant	Jograbi, GanhHI (h)	R	
39.		Pteroclidae	*Ptemcles exustus*	Indian Sandgrouse	Kal goial hakki (k) Kmr (h)	R	

40.	(Charadrifformes)	Scolopacidae	*Calidm minuta*	little Stint	Chfiota pankowwa (h)	W	
41.			GaHmago gallinago	Common Snipe	Ghana (h)	rW	
42.		Threskiornithidae	*Ptatalea teucorodia*	(Eurasian) Spoonbill	Chamach baza, Karanan	RW	
43.			*Pseudibs papullosa*	Black (Red-naped) ib:s	Kala baza, Karunuln	R	LRnl (IUCN)
44.			*Threskiomis aethiopica*	White(Black-headed) Ibis	Didhar, Safed baza (h)	R	
45.	Columbiformes	Columbrfae	*Chalcophaps indica*	Emerald Dcve		R	restricted distribution
46.			*Columba livia*	(Blue) Rock Pigeon	Kabutar (h)	R	
47.			*Ducula badia*	Maroon-backed Imperial-Pigeon		R	
48.			*Streptopelia chinesis*	Spotted Dave	China fakhta, Panduk (h)	R	
49.			*Streptopeta decaocto*	(Eurasian) Ring/ Collared Dove	Dhor fakhta, Parki, Gugn (h)	R	
50.			*Streptopelia tianquebarica*	Red Turtle Dove		R	
51.			*Treron phoenicoptera*	Common Green Pigeon	Hanal (h)	R	
52.	Coraciifomws	Aicedinidae	*Alcedo atthis*	(Common) small Blue Kingfisher	Chhota kilkila, Nita (h)		
53.		Cerylidae	*Ceryle rudis*	Pied Kingfisher	Koryala, Kilkila (h)	R	
54.		Dacetonidae	*Halcyon smyrnensis*	White-breasted Kingfisher	Kilkila, Kourilla (h)	R	
55.		(Alcedimdae)	*Pelargopsis capensis*	Brown-headed Strock-billed Kingfisher	Bada kii kiia (h)	R	
56.		Meropidae	*Merops leschenaulti*	Chestnut-headed Bee-eater	Lalsir patnnga (h)	R	
57.			*Merops orientalis*	Small Green Bee-eater	Patnnga Hamal (h)	R	
58.			*Merops philippmus*	Blue-tailed Bee-eater	Bada patnnga (h)	R	
59.	Cuculiformes	Centropodidae	*Ceniropus smensis*	(Greater) Coucal	Mahoka (h)	R	
60.		Cuculidae	*Cuculus micmplerus*	Indian Cuckoo			
61.			*Cuculus varius*	Common Hawk-Cuckoo	Kapak, Upak, Papiya (h)	R	
62.			*Camator (Oxyiophus)*	*Pied Crested Cuckoo*	*Kala papiya, Chatak (h)*	RS	

63.			Eudynamys scolopacea	(Asian) Koel	Koel (h)	R
64.	Gallijrmes	Phasiandiae	Cotumix-cotumix	Common or Grey Quail	Bada/Ghagus bhater(h)	RW
65.			Francolinus pondicerianus	Grey Partridge (Francolin)	Gowjal hakki (k)	R
66.			Galloperdix spadicea	Red Spurfowl	Chooti Jungli murgi (h)	R
67.			Gallus sonneratii	Grey jungli fowl	Kad koli (k)	R
68.			Pavo cristatus	Common (Indian) peafowl	Navilu (k) Mayura (h)	R
69.			Perdicula asiatica	Jungle Bush-Quail	Karilowya (k) Lowwan	R
70		Rallidae	Gallinula chloropus	Common (Indian) Morthen	Jal murghi, Pani murgh (h)	R
71.	Passeriformes	Alaudidae	Alauda gulgula	Indian small (Oriental) Skylark	Bharat (h)	R
72.			Eremopterix grisea	Blackbellied Finch-Lark	Diyora, Duri, Deoli (h)	R
73.			Galerida cristata	crested Lark	Ghandul (h)	R
74.		Cisticolidae	Cisticola juncidis	Streaked Fantial Warbler	Ghaki-phutki (h)	R
75.			Prinia sociallis	Ashy Wren Warbler	Phutki Kali phutki (h)	R
76.			Prinia subflave	Indian (Plain) Wren Warbler	Phutki (h)	R
77.			Artamus suscus	Ashy Swallow Shrike	Tadi ababeel (h)	
78.			Coracina (macei) novaehollandiae	Large Cuckoo-Shrike	Kasya (h)	
79.			Coracina melanoptera	Black-headed Cuckoo-Shrike	Jangli Kasy (h)	R
80.			Corvus macrorhyanchos	jungle Crow	Kala/Dhal kowwa (h)	r
81.			Corvus splendens	House Crow	Kowwa, Desi kowwa (h)	R
82.			Dendrocitta vagabunda	(Rufous) Tree Pie	Mahcarat (h)	R
83.		(Dicruidae)	Dicrurus caerulescens	White-bellied Drongo	Pahari bujanga, Dhapri (h)	R
84.		(Dicruidae)	Dicrurus adsimilis (D. macrocercus)	Black Drongo	Kola, Kotwal, Buchanga (h)	R
85.			Dicurus paradiseus	Racket-tailed Drongo	Bhimraj, Bharat (h)	R

86.		Oriolus oriolus	(Eurasian) Golden Oriole	Peelak (h)	R	
87.		Oriolus xanthoruns	Black (hooded) Oriole	Peelak, Zardak (h)	R	
88.	(Campephagidae)	Pericrocotus cinnamomeus	Small Minivet	Bulalchashm, Rajalal (h)	R	
89.	(Campephagidae)	Pericrocotus flammeus	Scarlet Minivet	Pahari bual-chashm (h)	R	
90.		Rhipidura aureola	White-browed Flycatcher		R	
91.		Tephrodomis Pondicerianus	Common Wood Shrike	Keroula (h)	R	
92.		Terpsiphone Paradisi	(Asian) Paradise Flycatcher	Shah bulbul, (h)	R	
93.	Fringillidae	Carpodacus erythirinus	Common Roserinch	Tuti Hal tuti (h)	rW	
94.		Emberiza bruniceps	Red (breasted) Bunting	Gandam (h)	W	
95.		Emberiz melanoephala	Black-headed Bunting	Gandam (h)	W	
96.	Hirundinidae	Hirundo concolor	Dusky Crag Martin	Chatan ababil (h)	R	
97.		Hirundo dauric	Red-rumped Swallow	Masjid-ababil (h)	R	
98.		Hirundo fluvicola	Indian Cliff Swallow	Nahar ababil (h)	R	
99.		Hirundo rupestris	Eurasian Crag-Martin		r	migrant
100.		Hirundo rustica	Common Swallow	Ababil (h)	RW	
101.		Hirundo smithii	Wire-tailed Swallow	Ababil (h)	R	
102.	Irenidae	Chloropsis aunfrons	Golden-fronted Chloropsis	Chota harrial Harewa (h)	R	
103.		Chloropsis cochinchinensis	Gold-mantled Chloropsis	Harewa (h)	R	
104.	Laniidae	Lanius schach	Rufous-backed Shrike	Mattiya/Kajala lafora (h)	R	
105.		Lanius vittatus	Bay-backed Shrike	Pachanak(h)	R	
106.	Muscicapidae	copsychus malabaricus	(white-rumped)	Shama (h)	R	
107.		Copsychus saularis	(Oriental) Magpie-Robin	Dhaiyal	R	
108.		Culicicapa ceylonensis	Grey-headed Flycatcher		r	
109.		Cyornis tickellae	Tickel's Blue Flycatcher		R	
110.		Monticola cinclorhynchus	Blue-headed Rock Thrush		r	
111.		Moniticola solitarius	Blue Rock Thrush		rW	

112.		Muscicapa thalassina	Verditer Flycatcher		r
113.		Myiophonus horsfieldii	Malabar Whistling-Thrush	Gunta ukkee (k); Kasturo (h)	R
114.		Saxicola caprata	Pied Bush Chat		R
115.		Saxicola caprata	Collared Bush Chat		??
116.		Saxicola, ioes fulicata	Indian Robin	Kalchuri (h)	R
117.		Turdus merula	Eurasian Blackbird		R
118.	Nectariniidae	Aethopyga siparaja	Yellow-backed Sunbird		r
119.		Dicaeum agile	Thick-billed Flowerpecker		R
120.		Dicaeum erythrorhynochos	Tickell's Flowerpecker	Phoolchuki (h)	R
121.		Dicaeum kignipectus	Fire-breasted	Phoolchuki (h)	r
122.		Nectarubua asiatica,	Purple Sunbird	Phul soongi (g)	R
123.		Nectaninia zeylonica	Purple-rumped Sunbird	Shakarkhora(h)	R
124.	Paridae	Parus major	Grey (Great) Tit	Ramgangra (h)	R
125.		Parus xanthogenys	Yellow-cheeked Tit		r
126.	Passeridae	Anthus novaesslandiae	(Indian) Paddyfield	Rugail, Charchari (h)	R
127.		Lonchura malabarica	White-throated Munia	Charga, Pidda (h)	R
128.		Lonchura striate	White-backed Munia		R
129.		Lonchura punctulata	Spotted Munia Pipit	Telai munia Seenabaz (h)	R
130.		Motacilla alba	White Wagtail	Dhoban (h)	rW
131.		Motacilla cinerae	Grey Wagtail		rW
132.		Motacilla Citreola,	Yellow-headed Wagtail	Pilkya (h)	rW
133.		Motacilla, flava	Yellow Wagtail	Pilkya (h)	W
134.		Motacilla madaraspatensis	Large Pied Wagtail	Mamula, Khanjan (h)	R
135.		Passer domesticus	House Sparrow	Chusi, Khas churi (h)	R
136.		Petronia xanthocollis	Yellow-throated sparrow	Raji, jangli china (h)	R
137.	(Ploceidae)	Ploceus manyar	Streaked Weaver Bird	Bamani baya (h)	R
138.	(Ploceidae)	Ploceus philippinus	Baya Weaver Bird	Baya Son-chin(h)	R
139.	Pattidae	Pitta brachyura	Indian Pitta	Naurang (h)	R
140.	Pycnonotidae	Hypsipetes indicus	Yellow-browed Bulbul		r
141.		Pycnonotus cafer	Red-vented Bulbul	Guldum (h)	R

142.			Pycnonotus jocosus	Red-whiskered Bulbul	Kamera bulbul (h)	R
143.			Pycnonotus luteolus	White-browed Bulbul		R
144.		Sittidae	Sitta castanea	Chestnul-bellied Nuthatch	Siri Katphoriya (h)	R
145.			Sitta frontail	Velvet-fronted Nuthatch		R
146.		Sturnidae	Acridotheres fristis	(Common) Indian Myna	Gorwantera (k); Desi myna (h)	R
147.			Groacula religiosa	Hill Myna	Pahari myna (h)	R
148.			Sturmus malabaricus	Grey-headed Myna	Powei (h)	R
149.			Sturmus pagodarum	Brahminy Myna (Starting)	Popoya myna, Puhaia (h)	R
150.		Sylviidae	Chrysomma sinense	Yellow-eyed Babbler	Bulbul/Gulab chashm (h)	R
151.			Dumetia hyperythra	Rufous-bellied Babber		R
152.			Orthotomus sutorius	Common Tailorbird	Darzee Phutki, Piddi (h)	R
153.			Pellomeum ruflceps	Spotted Babbler		R
154.			Pomatorhinus schisticeps	White-browed Scimitar		r
155.			Turdoides caudatus	Common Babbler	Ganga, Dumri, Chilchil sor (h)	R
156.			Turdoides makolmi	Large Grey Babbler	Gongya (k); Ghogi, (h)	R
157.			Turdoides striatus	Jungle Babbler	Sat bhai (h)	R
158.		Zosteropidae	Zosterops palpebrosus	Oriental White-eye	Baboona(h)	R
159.	Piciformes	Capitonidae (Megalaimidae)	Megalaima haemacephala	Coppersmith or Crimson-breasted Barbet	Tambayat, Basanth lisora, Katphora (h)	R
160.			Megalaima zeylanica	Large Green Barebt	Bade basanta (h)	R
161.		Picidae	Dinopium benghalense**	Lesser Golden-backed Woodpecker		
162.			Drycopus javensis**	Great Black Woodpecker		
163.			Microptemus (Celeus) brachymus	Rufous Woodpecker		
164.			Picoides (Dendrocopos) mahrattensis	Mahratta or Yellow-fronted Pied Woodpecker	Katphora, (h)	R

165.			Picoides (Dendrocopos) nanus	Pigmy (Brown-capped) Woodpecker		R
166.	Pesittaciformes	Psittacidae	Psittacula cyanocephala	Blossom-headed Parakeet	Tuiya tota (h)	R
167.			Psittacula eupatria	Alexandrine Parakeet	Hirama-/Rai-tota	R
168.			Psittacula Krameri	Rose-ringed parakeet	Tota, Lyber tota (h)	R
169.	Strigiformes	Caprimulgidae	Caprimulgus asiaticus	Common (Indian) Nightjar	Cahpka (h)	R
170.		Strigidae	Bubo bubo	Indian Great-homed (Eurasian Eagle) Owl	Ghughu (h)	r
171.			Bubo (Ketupa)	Brown Fish-Owl	Ullu, Amrai-ka-ghughu (h)	R
172.		Tytonidae	Tyto alba Zeylonensis	Barn Owl	Kuraya, Karail, Buri churi (h)	R
173.	Trogoniformes	Trogonidae	Harpactes fasciatus	Southern (Malabar) Trogon	Kakarne hakki (k), Kafni	R
174.	Turniciformes	Tumicidae	Tumix suscitator	Common or Bluelegged	Karechakki (k); Guru	R
175.	Upupiformes	Upupidae	Upupa epops	Eurasian Hoopoe	Hud hud (h)	RW

**Unsure as to whether this is the species recorded deduced according to description given in MP2.

Mammalian Populations

Protected areas of Karnataka have remarkably a complete assemblage of predators, larger herbivores and lesser known mammals. Predators include, the tiger *Panthera tigris* the leopard *Panthera pardtts* and dhole or the asiatic wild dog *Cuon alpinus* and the wolf *Qntis lupus*. The larger herbivores include seven species of ungulates; the Asian elephant *Ekphus niaximiis* the gaur *Bos gnurus*, the sambar *Cervus unicolor*, the chital *Axis axis* the muntjac *Muntiacus muntjak* and the four horned antelope *Tetracms quadricornis*, the black buck *Antelope cervicapra*. The primates include: the hanuman langur *Presbytis entellus*, the liontailed monkey *Macaca silenus* and bonnet monkey *Macaca radiata* also use the same niche. The omnivores such as the sloth bear, *Meliirsus ursiniis* and the wild pig *Sus scrofia* also use the same level ground for their existence.

The lesser known mammals include: the slender loris Loris tardigradus, the jungle cat *Felis chaus*, the leopard cat *Felis bengalensis*, the common mongoose *Herpestes edwardsi*, the striped necked mongoose *Herpestes vitticollis* the small Indian civet *Viverricula indica*, the mouse-deer *Tragulus meminna*, the black naped hare *lepus nigricollis*, the Indian giant squirrel *Ratufa indica*, the Indian porcupine *Hystrix indica* and the Indian pangolin or scaly ant eater *Manis crassicaudata* use the same habitats.

Herbivores

Protected areas of Karnataka are rich in herbivore populations, apart from large number of predators. Very little study is made on species diversity and their ecology.

The Elephant (Elephas maximus L)

Generally only males have large tusks. Rarely the tusks of females (called as tushes) scarcely protrude or perhaps protrude a few inches. Some males do not possess tusks and are known as makhnas. The makhnas are large in build with extraordinary well developed trunk to compensate the absence of tusks. Herds are believed to be composed of single families and controlled by the oldest female, called matriarch. Adult males are longer, visiting herd to herd in search of females in estrus. The adult males, rarely adult females are subject to peculiar periodical paroxysms of excitement, called musth. The elephant is a dominant mammal both in terms of numbers and biomass. It is distributed throughout the stretch of Western Ghats and Eastern Ghats including foothills of Nilgiris that connects both. The number of elephants estimated are 600-800 at Nagarahole reserve and 800-1000 at Bandipur in the area of 643 sq. km and 880 sq. km, respectively, depending on the seasonal variation. The density works out at the rate of 1 elephant per sq. km. In other forests of the state the density varies from 0.1 to 0.5 per sq. km.

The gaur or Indian Bison (Bos gaums)

With its huge head, deep massive body and sturdy limbs, the Gaur is the embodiment of vigor and strength. A very striking in the male Gaur is the muscular ridge upon the shoulders which slopes down to the middle of the back where it ends in an abrupt dip. A newly-born gaur is a light golden-yellow, which soon changes to light brown and so to coffee or reddish brown. The old bulls are black, their bodies are almost hairless with an ashy forehead and yellowish or white feet.

This wild ox is the second largest mammal of the forests. Gaurs are mostly seen in the moist localities, such as the tall grasslands in wet evergreen and deciduous forests. Herds of 5-30 gaurs, in addition to solitary bulls are sighted frequently in the protected area. These are more commonly seen in large herds during wet season. Based on the season and sightings the average gaur population is estimated to be 1800 gaurs at Nagarahole and 1500 gaurs at Bandipur, giving an average density of 3 and 2 gaurs/sq.km respectively. Density is undoubtedly higher at Nagarahole than at Bandipur. The gaur population was severely affected by the RPD epidemics during 1967 and 1969. The species took nearly 30 years to reach steady population. Gaurs are primarily selective grazers, but also browse on members of the Malvaceae. They occasionally raid the crops in the adjoining fields.

The Sambar (Cervus unicolor Kerr)

The sambar is the largest and graceful Indian deer and carries the horns, called antlers. The antlers are stout and rugged, and they are shed periodically. The Sambar is widely distributed through all the habitat types, though solitary animals or in small groups are commonly seen. On occasions they are found in large gatherings up to 18-24 when food supply is plenty. Red patch at the neck, that appears seasonally, in both the sexes is connected with the breeding dress. No objective assessment of their numbers is made owing to thick understorey growth of vegetation. It is estimated that the population is 2000-2500 at Nagarahole and 1500-2000 at Bandipur, accounting for 4 and 2 sambar per sq. km respectively. They prefer dense forests, tall grass areas and take shelter in thick undergrowth. Sambar being shy, infrequently seen, until such time, that it comes out in the open on its way to water hole or salt-licks. They move in a narrow home ranges of their own, marking the boundary by urination and repeatedly dung droppings at the particular places. Its alarm call echoes in that dense forest canopy, on sighting/spotting a danger, stamping its fore-foot. All the denizens of the jungle get alerted against impending danger, by its warning calls.

The Chital (Axix axis E.)

The chital is perhaps the most beautiful of all deers. Its coat is bright rufousbrown profusely spotted with white at all ages and in all seasons. Old bucks are more brownish in colour and darker. One always associates chital with beautiful scenery with grassy forest glades and shaded streams. Chital are seen more commonly in large herds in open glades of tourism zones, open scrub and dry deciduous forests. They are found in smaller herds in moist deciduous forests. They avoid very dense forest. The population of chital in both the reserves is quite high. They occur in sizable herds especially in open grasslands. They are found freely wandering about near forest guest houses and forest staff quarters, where there are openings in the forest canopy. The population may vary between 3000-3500 at Nagarahole and 2800-3200 at Bandipur, accounting for 5 and 4 chital per sq. km respectively.

The muntjac (Muntiacus muntjac Zimmerman)

When alarmed and in flight the deer give out a series of short cackling barks. A sharp rattle likened to the sound of castanets heard when muntjac are bounding away is a vocal noise. Muntjac are seen usually as solitaries or in pairs. Being shy animal very occasionally seen. It is found in dense forests, thick undergrowth, moving within their limited home range. Home ranges are marked by urination and heaps of dung. It is a difficult animal to count, being small and inhabiting thick hushes. They return to their hiding place at the slightest sound giving a alarm call like that of barking of a dog. The barking deer is the most difficult animal to estimate the population size. However a forest official estimate may reveal the density of the animal at 4 animals per sq. km. both in Nagarahole and Bandipur.

Four Horned Antelop Or Chousingha (Tetracerus quadricornis Blainvelle)

The fourhorned antelope or Chowsingha is the only member of this group of antelopes, with two pairs of horns. Of these front pair are always the shorter. One of the most interesting features in the chowsingha is the presence of a pair of well-developed glands between the false hooves of the hind legs in both males and females. Chowsingha are usually seen alone or in pairs. They have a habit of defecating at one place. Chowsingha lives in tall grassy blankets of undulating or hilly country or open jungle. The animal is very difficult to count as it bounds off at the slightest noise and generally keeps safe distance from humans.

The Common Langur or Hanuman Monkey (Presbytis entellus Dufreshne)

This is the long-limbed, long-tailed, black-faced monkey, seen around towns and villages and in forests as well. Langurs are more arboreal in habit than macaques. Forest dwellers are more shy and wary, particularly if hunted by forest tribes. Langurs are pure herbivores, they eat wild fruits, flowers, buds, shoots and leaves. Some times they eat only petioles and discard leaves, which is evidenced by the forest floor covered with discarded leaves. Other herbivores such as elephants, sambar, chital and even humans benefit from the fruits fallen on the forest floor. The vertebrate enemy of the langur is the panther. The sight of one or of a tiger or of any animal, including human that arouses suspicion, produces the guttural alarm note which sends the whole troop bolting. Other denizens of the forest get alerted from the alarm calls as to the prowling beasts of prey in the vicinity. The population size in Nagarahole and Bandipur reserves may be around 6 and 4 animals per sq. km. respectively.

Bonnet Monkey (Macaca radiata Geoffroy)

A medium-sized, long-tailed macaques. A bonnet of long dark hairs radiates in all directions from a whorl on its crown. The bonnet does not quite cover the forehead, where the hairs are short and neatly parted in the centre. It is common species in village and jungle and is more arboreal in habit than most macaques, hence it possess longer tail. Bonnet macaques eat anything eatable, fruits, berries, leaves and shoots, also insects, grubs and spiders. Field studies show that bonnet macaques are very social and live in highly organised troops of both sexes of all ages in well defined territories, which may vary from 3-5 sq. km. Social life is based on a well marked dominance hierarchy among the males. The population density being very thin in Nagarahole and Bandipur may vary between 0.1 to 0.2 animals per sq. km.

The Black Buck or Indian Antelope (Antelope cervicapra L)

The Blackbuck, that once roamed the fringes of Nagarahole and Bandipur reserves, is the sole representative in India of the genus Antelope. Its striking colour and its beautiful spiraled horns, gives it an elegance hardly equaled by any antelope. This exclusively Indian animal is perhaps the most beautiful of all its kind. The Blackbuck occurs as an isolated population, practically, in plains and avoid forests or hilly areas. It is usually seen in herds of 20 or 30 in parts of Mysore, Mandya, Dharwad, Bellary and Bidar Districts of the state. In other areas it is hunted out. The population of Blackbuck ranges around 3000-4000 in the entire state.

The Lion-tailed Monkey (Macaca silenus L)

It is distinguished from all other species of macaques, by long dark grey or brownish grey hairs growing from the temples and cheeks and also by its glossy black coat, occurs as isolated populations in tropical wet evergreen forests in Uttara Kannada and Dakshina Kannada Districts. Like other macaques these monkeys are gregarious living in troops of 12-20 animals or more. The call of the male is said to resemble the human voice. It is compared to the coyeh of a man trying to get in touch with his lost companions in the jungle, and again to the loud coo of the pigeon. Their movements through the tree tops are usually slow and deliberate. Their exact population is not known and is one of the most endangered monkey of India.

Omnivores

The Indian Wild Boar or Wild Pig (Stis scrofa L)

Indian wild Boar lives in grass or scanty bush jungle including thick forests. They are omnivorous feed on crops, roots, tubers, insects, snakes, offal and carrion. No animal is more destructive to crops and in cultivated fields than these of wild boars. Wild boars display great intelligence and few animals show greater courage and determination. The sense of smell is acute and eyesight and hearing moderate. Wild boars are highly prolific. They apparently breed at all seasons.

The wild pig is widely distributed in the forests. After the elephant it is the most serious mammalian pest of cultivated crops. No population estimation is available, as it is difficult to count them. The animal makes a favorite menu for tigers. They are found in all types of forests eating anything and everything. Being an omnivorous, it feeds on vegetation roots of grasses and some specific roots and tubers, even on carcass and velvety horns of chital.

The Sloth Bear (Melursus ursinus, Shaw)

With its elongated muzzle and lower lip, long unkempt hair and short bind legs, sloth bear is the most uncouth of all bears. Most have whitish V-shaped breast patch and usually the muzzle and the tips of the feet are dirty white or yellowish. The claws, always longer on the forefoot and are ivory white. Sloth bears live where there is sufficient food and favourable places, where out cropping of rock and tumbled boulders offer them shelter during the odd weather and the rains. They are known to occupy abandoned buildings and temples in the forest.

The bear is wide spread over all the habitat types of the forests, though at a very low density, say 0.2 to 0.4 animal per sq. km. The food consists mainly of fruits, flowers, insects, but a hungry sloth bear may be driven to eat carrion. The great combs of the large rock bee *Apis dorsata,* suspended in clusters under rocks or from a branch and the combs of smaller forest bee *Apis indica* usually hidden in tree hollows, that are loaded with honey during hot weather when wild figs, mangoes, jamoon, bael and ebony trees are in fruits. The bear climbs for fruits or shakes it down with its great paws. It knocks down a honeycomb and descends to feed on it. In the monsoon insect food is more plentiful and bears can find many insects under stones, under fallen logs, under bark and crevices of trees. But their main insect food item is termites.

Carnivores

The Tiger (Panthera tigris L)

The Indian tiger is a rich-coloured, well-striped animal with a short coat and great individual variation. Ordinarily, the tiger hunts between dusk and dawn. It hunts game animals of all kinds including young ones of elephants and gaurs. It preys on deer, sambar, wild boar, bears and porcupines. It also kills and eats panthers and other tigers (cannibalism). Driven by hunger it will eat almost anything, fowl, fish reptiles, and carrion. The tiger is a highly territorial animal leading a solitary life and pairing at a time when female comes in heat.

The tiger is highly endangered species and is entered in the "Red Data Book". Tiger is placed at the apex of the biological food pyramid and controls the entire ecosystem. The implementation of "Project Tiger" – in the country during 1973 saw the rejuvenation of the species. The tiger was indiscriminately hunted out to low population of 11 at Bandipur and 6 at Nagarahole during 1974, when they were declared as national parks.

The Leopard or Panther (Panthera pardus L)

A typical panther is a sleek short-haired animal with a fulvous or bright fulvous coat marked with close-set black rosettes. Panthers are able to live and thrive almost anywhere. They are not restricted to forests or heavy cover like tigers, and thrive as well in open country as among rocks and scrub jungle. The panther will kill and eat anything it can overpower with safety, cattle, deer, monkeys, the smaller beasts of prey and larger rodents like porcupines.

The strength of the panther is amazing, while it scrambles up a tree carrying a full grown chital stag in its mouth to find security for its kill. The leopard is an associate carnivore, that hunts on a ground level with tiger and wild dog in the forests, but it preys on smaller herbivores. This predator prefers shrubby jungle, frequently straying out into villages to lift dogs and livestock. No reliable estimates of leopard population are available.

The Dhole or Wild Dog (Cuoan alpinus Pallas)

It is much like a domestic dog in general appearance, with long and lank body of the wolf, but relatively shorter in leg and muzzle (long tapering mouth). The dhole usually has 12 to 14 teats as against ten in domestic dogs (Cam's). Very distinctive is its red coat with black bushy tail. The pack is a family, some tunes of several families. Such union increases the chances of killing larger prey. Usually wild dogs hunt by day, rarely by night. Panthers and bear and even tigers have been attacked and killed by wild dogs. The wild dogs hunt on the same species of prey animals on which leopards and tigers hunt. The pack of 916 animals are common in the forests. These are aggressive hunters and considered as effective predators to weed out the sick, week and aged, thus help the parks to be healthy. Unfortunately their population is rapidly dwindling and the correct population density is not known.

Scavengers : The jackal (Canis attreus L)

The jackal's long-drawn eerie howling at dusk or before dawn is perhaps more familiar to most people, camping in the forest guest houses, than the animal itself. Its nearest wild relative is the wolf, but the jackal is smaller in build and meaner in aspect. The greater number live in the low lands, about towns and villages and cultivation, sheltering in burrows among ruins and in dense grass and scrub, jackals usually come out at dusk and retire by dawn, but in cool and cloudy weather they may come out. They do good work in the clearance of carcasses and offal, sharing with vultures, the only sanitary service known to many of our towns and villages. The hunting instinct is not wholly dormant in the jackal, some become poultry thieves or take to killing of lambs and sick goats. The jackal, that used to be a common scavenger is hunted regularly by a nomadic tribes for their skin.

A dog-like build, massive head with powerful jaws that can crush big bones, weak hindquarters and a heavy dorsal crest of lung hairs, sharply defined from rest of the coat, distinguish the hyena. Its colour varies from cream buff or tawny to the grey or dirty white of the harsh scanty summer coat. Transverse stripes on body and limbs usually well defined. The hyena is rare in forested areas, abundant in open country, especially, where low hills and ravines offer convenient holes and caves for shelter. Animals that have died of disease or those killed by the larger beasts of prey are the usual food of the hyena. Occasionally they form groups, called packs and hunt even larger animals. Sometimes they turn out to be child lifters. It is almost hunted out in the state.

Table 23.11. Mammlian diversity of Karnataka State. (Prater, 1998: Gina, 2000)

No.	Order	Family	Species	English name	Local Name (Kannuda)	Distribution in India	Status	Status I	Status II	Status III
1.	Artiodactyla	Bovidae	Bos fronyalis (B. gaurus)	Gaur	Kadu yethu, Kartee	L	VU Aled+2cd, Cl +2a (IUCN)	A	B	A
2.			Tetracercus quadricornis	Four-horned Antelope	Koondoo kooree Poki		VUC2a (IUCN)	C	D	D
3.		Cervidae	Axis axis	Chital (Spotted deer)	Saranga Jinke	W	S	A	B	A
4.			Cervus unicolor	Sambar	Kudawe, Kaddama	W	S	B	B	B
5.			Muntiacus muntak	Muntjac (Barking deer)	kard koorie	W	S	B	C	B
6.		Suidae	Sus scrofa	Wild boar	Hundi		W	S	A	A
A										
7.	Tragulidae	Moschiola	Mouse deer meminna	Koor andi, Kooray (Indian Chevrotain)		W	R			
8.	Carnivora	Canidae	Canis aureus	Jackal	Nuree, Kunni nuree			A	A	A
9.			Cuon alpinus	Asiatic Wild-dog (Dhole)	Kadu nai, Chira nai		VUC2a (IUCN)	A	B	C
10.		Felidae	Panthera pardus (Panther)	Leopard Siwangi, Keeraba	Chirchu, Chirate,			D	D	D
11.			Panthera tigris	Tiger	Hoolee		EN A2cd (IUCN)	B	C	C
12.			Prionailurus (Feils) (bengalensis)	Leopard cat						
13.		Herpestidae	Herpestes	Common	Keeree					

			edwardsii	mongoose						
14.			Herpestes fuscus	Brown Mongoose						
15.			Herpestes vitticoilis	Stripe-necked mongoose	Kemp keeree					
16.		Hyaenidae	Hyaena hyaena	(Striped) Hyaena	Kathe keeruba, Kathe kurbua, Nai Hooli					
17.		Mystelidae	Lutra lutra	European/ Common otter	Neeru nai, Uddra					
18.		Ursidae	Malursus	Sloth bear Ursinus	Karadee		VU A2cd, Cl+2a(IUCN)	D	B	C
19.		Viverridae	Paradoxurus hermaphroditus	Common palm civet	Mara bekku, Mantta					
20.			Viverra	Malabar civet Megaspila			Nearing extinction!			
21.			Vicerricula indica	Small Indian civet	Poonagubekku, poonugu kotthi					
22.	Lagomorpha	Leporidae	lepus	Indian hare Nigricollis	Mola	W				
23.	Phohdota	Manidae	Manis crassicaudata	Pangolin			LR/nt (IUCN)			
24.	Primates	Cercopi-dae	Semnopithecus entellus (Hanuman monkey)	Comon langru	Moosoo, Moosoowa	W	LR/nt (IUCN)	A	A	A
25.			Macaca radiata	Bonnet				D	C	C
26.		Loridae	Loris tardigradus	Slender loris	Mala manushya, Kadapapa, Adavi manushya	(IUCN)	VU Alcd			
27.	Proboscidea	Elephantidae	Elephas	Asian elephant	Ane	W	EnAlcd	B	B	B

				Maximus			(IUCN)			
28.	Rodentia	Hystrix indica	Indian	Yed Moolu handi porcepine (Common)		W		B	B	B
29.		Pteromyidae	Petaurista	Giant Flying squirrel						
30.		Sciuridae	Funambulus	Three striped palmarum	palm squirrel					
31.				ratufa indica (Indian) gaint	Malabar		VU B1 + 2c (IUCN)	A	A	B

Note: I = Mutnodi; II = Hebbe; III = Lakkaralli

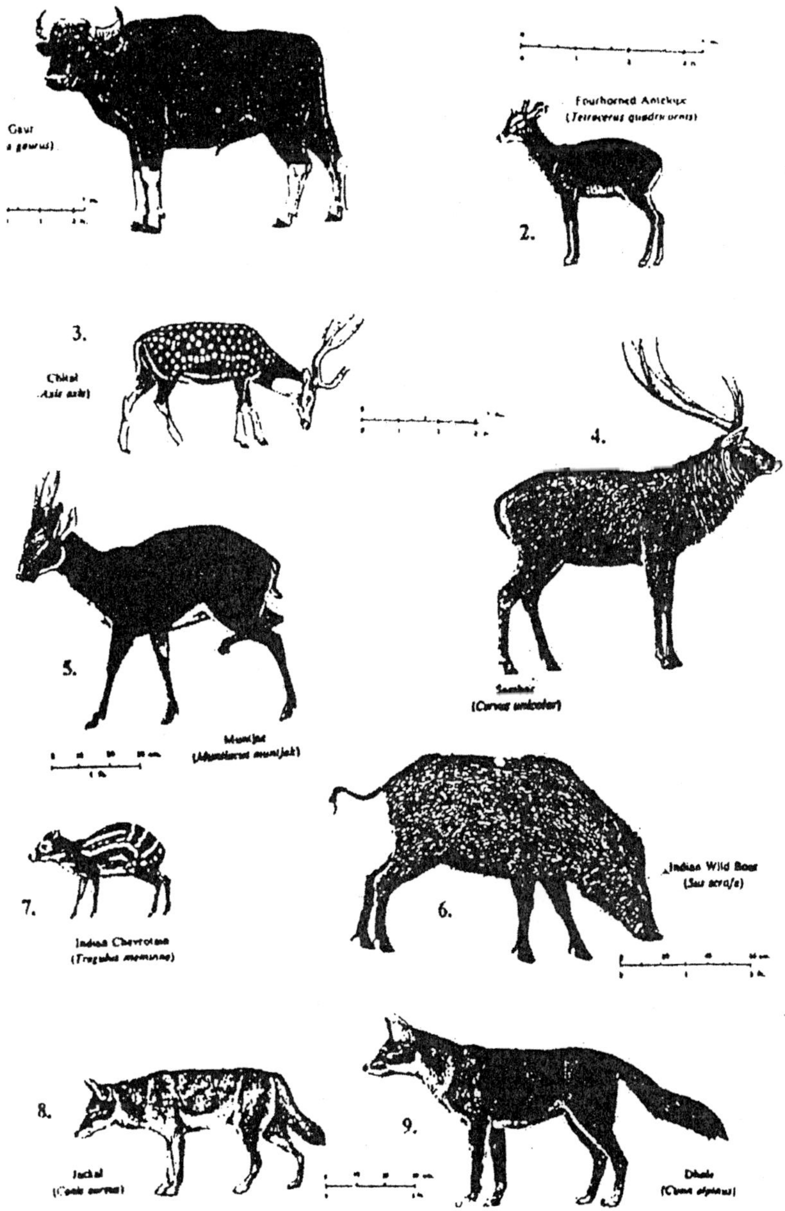

Fig. 23.3 : Mammalian diversity of Karnataka State.

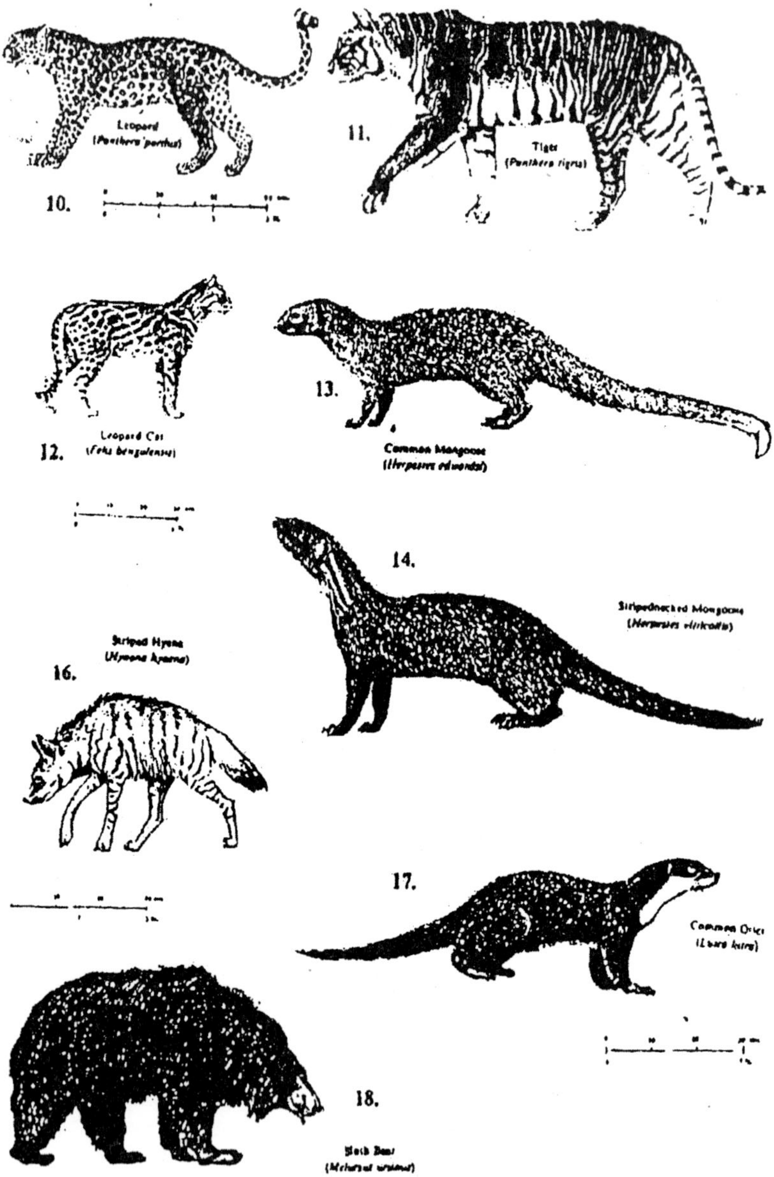

Fig. 23.3 Continued

Fig. 23.3 Continued

Ms. Gina Caplen, University of Salford, England, travelled all the way to India to study the bio-diversity of mammals and birds of Karnataka. She stayed in the B.R. Project Kuvempu University campus for more than four months and reported 31 types of mammals and are shown in Table 23.11 and Plate 23.1.

Mangroves

Mangroves are salt tolerant coastal forest ecosystems found in the tropical, subtropical and intertidal regions of the world. Mangroves are the reservoirs of various species of plants, microbes and animals associated together over a long evolutionary time and still these are imperfectly known. The early study on Indian mangroves dates back to 1814. That is Roxburgh obtained specimens from Sundarbans through Buchanan Hamilton and William Care during 1814 and described the characteristics of the plants in his book Hortus Bengalensis. Hooker's book, Flora of India also explains the mangroves of India. The subsequent contribution in the early part of 20th century include Hooker, 1872 : Blatter, 1905, Cooke, 1905 : Gamble, 1936: Navalkar, 1951.

These are complex ecosystems, which are influenced by tropical rains, tidal interactions, wind, cyclone and different habitat conditions. Mangrove swamps are called Mangal bans (Macnae, 1968). They cover 60-75% of the tropical and subtropical coastal ones. Mangroves are a cluster of tropical plant-formations growing along the land-sea surface and in estuaries and lagoons, reaching upstream in the rivers upto a point where the water still has some salinity.

Mangroves occupy a highly strategic position in the economy and the ecology of the coastal areas of countries where they occur. Economically they have been important primarily as aquaculture or fish pend sites, as source of forest products, as seed gathering grounds, etc. Ecologically, they are important in near shore nutrient enrichment, as spawning, breeding, feeding and nursery grounds for economically important species of aquatic life, in shoreline stabilization and protection.

However, rapidly increasing human population over the years has necessitated the extensification of the utilization of mangroves. The activities related to that of fish ponds, agriculture and development activities led to the degradation of the mangrove forests and near shore ecosystems.

Mangroves are very important because they possess enormous bio-diversity and provide biologically most productive habitat. Mangroves cover an area of about 17 million ha, in the whole world. The total area of mangrove in India is estimated to be 6740 Sq. km (Table 23.12), which constitutes 7% of the world mangroves and 8% of the Indian coastal line (7600 km) (Status Report, 1987). Indian mangroves comprise 30 genera and 62 species categorized in 21 families (Blasco, 1975).

It is estimated that Karnataka has a vast area of 5000 ha in the Dakshina Kannada district and 3000 ha Uttara Kannada District. Mangroves of Karnataka are of the fringing type found in the intertidal region along the estuaries, backwaters and islands. The distribution of mangrove vegetation depends on the intertidal expanse, tidal range, substratum, salinity of water and soil.

Karnataka possess coastal stretch of about 300 km. It provides rich economic returns from quality fishery resources of prawn, fishes and others. Being tropical, the rate of production of mangroves (litter and biomass) exceptionally high and hence it exports nutritive fractions of decomposing litter to the estuaries and coastal waters. This phenomenon is enhanced by the tidal action. In estuaries and sea, it supports the good growth of phytoplankton followed by zooplankton-food chain and food webs efficient transfer of energy in the ecosystem.

Table 23.12. Area wise Distribution of Mangroves Forests in India

	States/Union Territory	*Area in sq. km (Values round to the nearest 10 sq. km)*
1.	Andaman and Nicobar groups (including Nicobar islands)	1,190
2.	West Bengal (the Sunderbans)	4,200
3.	Orissa (All deltaic and coastal mangroves, including the Mahanadi area.)	150
4.	Andhra Pradesh (the Godavari- and the Krishna.)	200
5.	Tamil Nadu (the Cauvery and 150 adjacent coastal stretch.)	
6.	Karnataka (Coondapur and Malpe karwar Patches.)	60
7.	Goa	200
8.	Maharashtra (inclusive or Ratnagiri, Vijayadurg, Malvan, Devgad, etc. inclusive of 85 sq. km.)	330
9.	Gujarat (including the Narmada- the Tapti-Gulf of Khamphayat mangroves of 105 sq. km.)	260
	Total	6,740

Source: Mangroves of India-status report 1987.

Species diversity found along the coasts of Karnataka has been listed below. And the taxonomic groups have been showed in Table 23.13.

1. *Rhizophora mucronata*
2. *Rhizophora conjitgata*
3. *Rhizophora apiculata*
4. *Avicennia officinalis*
5. *Avicennia alba*

6. *Sonneratia caseolaris*
7. *Sonneratia alba*
8. *Acanthus ilicifolhts*
9. *Kandelia candel*
10. *Derris tnfoliata*
11. *Derris scandcns*
12. *Ckrodendrnm incnnc*
13. *Brugniera gynmorrhiza*
14. *Bruguiera cylindrica*
15. *Exeoecaria agn1locha*
16. *Ceriops decandra*
17. *Aegiceros corniciilatam*
18. *Cerebera odolhim*
19. *Lumnitzera racemosa*
20. *Dalbergia spinosa*
21. *Acrostichum auraim (Perm)*
22. *Protersia coractata*

The important mangrove areas of the Karnataka coast are found near the rivermouth areas of the various rivers. They have been listed below:

Mangalore and Udipi Districts

1. Nethravati River estuarine area-Jappina Muguru (Mod).
2. Gurpur River estuarine area-Kodiyalbail, Bengra Kulur, Padukodi (All Good), Boluru (Mod).
3. Mulki River estuarine area-Mulki, Hejjamadi Kodi, Kokrani, Hejjamadi, Chitrapu*, Sashihitlu, Koluvayalu, Haleyangadi and Pavanje (All VG).
4. Udiyavara Hole estuarine area-Katapadi, Yenagudde and jarududru (Mod), Udiyavara, Kodekar (Good), Mattu*, Pitrodi (All very good).

5. Swarna-Sita estuarine complex : Haradi, Hangarakatta, Airodi parampalli, Kotattatu-(very good).
6. Chakra-Halladi-Kollur estuarine complex: Vodara Hobli, Anegundi, Jaladi, Kamastara Kudru*, Uppinakudru, Bantwadi, Hadavu-(all very good), Kodi Heri Kudru, Senapur-(all good).
7. Baindur Hole: Paduvari*, Uppunda-(very good).
8. Shiroor Hole-Shiroor-(very good).

Kanvar District

9. Venkatapur River estuary: Heble-(very good) Alvekodi, Shirali (good).
10. Sharavati and Bedgane estuarine complex : Hosapatna (good), Mavina Kurve (poor), Pavina Kuruve (Very poor).
11. Aghanashini River estuarine complex : Kimmani, Gudkogal, Kagal, Devenagi Mudangi, Sanikatta posakatta-(all good), pini (good), Gudeangadi-(Moderate).
12. Gangavali and Belekeri Estuarine complex: Gangavalli, Shiroor (poor), Bilihongi, Juga-(very good), Sagad Siri (good).
13. Kalinadi and Mavinahalli Estuarine Complex : Cunjkeri, Kadavad, Siddar, Hanklon (Very good), Hagarishitta, Madhevade, Vilasbandar (good), Kanasgiri, Sadashivganh (Moderate), Siddapur, Ambejug, Bad Jug, Basava Naga Hankonjung, Devbeg (all Very good).

Rating	-	(10 × 10 mt. Plot)
Very good (VG)	-	More than 50 trees
Good (G)	-	36-50 trees
Moderate (M)	-	16-35 trees
Poor (P)	-	6-15 trees
Very poor (VP)	-	less than 5 trees
Other sites	-	are prospective area

Table 23.13. Species diversity of principal mangrove vegetation of India

	Family	*Genera*	*Total No. of species known*	*Total No. of species known from Indian Ocean, Pacific Region and (Indo-W. Pacific)*	*Number of Species in India*	*Remarks*
1.	Rhizophoraceae	1. Rhizophora	7	5	4	*Rt. stylosa* (A & B islands and Pichavaram) West Coast
		2. *Brugiera*	6	6	4	*R. lamarckii* known from Pichavaram
		3. *Ceriops*	2	2	2	
		4. *Kandalia*		1	1	*K. candel. K. rheedi* occur mainly in Goa-Ratnagiri (Described as B. coccinea in Nicobar)
2.	Avicennianceae	5. Avicennia	11	6	3	
3.	Combretaceae	6. Lumnitzera	2	2	1	
4.	Sonneratiaceae	7. Cymnometra		1	1	
		8. Sonneratia				
5.	Rubiaceae	9. *Schyphiphora*	5	5	1	S. *caseolaris* known only from Goa
6.	Plumbagianceae	10. *W. Aequitilis*	2	2	1	

7.	Meliaceae	11. *Xylocarpus*	10	8	4	Southern Limit: Pichavaram (Tamil Nadu). West Coast
8.	Myrsinaceae	12. *Aegiceras*	2	2	1	
9.	Acanthaceae	13. *Acanthus*	3	3	2	
10.	Euphorbiaceae	14. *Excoecaria*	1	1	1	
11.	Papilionaceae	15. *Dalbergia*	2	2	2	
		16. Derris	2	2	2	
12.	Palmae	17. *Nypa*	1	1	1	
		18. *Phoenix*	2	2	2	
13.	Sterculicceae	19. *Heritiera*	2	2	2	
14.	Bombacaeae	20. *Camptostemon*	2	2	2	Recorded less from East Coast also (Orissa) now confined to the West Coast.
15.	Salvaqoracede	21. *Salvadora*			1	
16.	Apocyanaceae	22. *Cerbera*	2			
17.	Aizoaceae	23. *Sesuvium*			1	
18.	Chenopodiaceae	24. *Arthicbemum*			1	
19.	Polypodiaceae	25. *Salicornid*			1	
		26. Suaeda			3	
20.	Filicices (Pteridophyta)	27. *Acrostichum*			1	
21.	Poceae	28. *Porterasia*			1	
		29. *Aeluropus*			1	
		30. *Urochondra*			1	

Source: Mangroves of India-status report. 1987.

Economic Values

These form a unique ecosystem comprise land, water, vegetation, animals and also shelter to human beings. The natural combination results in the production of multiplicity of goods and services. The following are the important economic aspects of mangroves:

- Source of timber
- Minor forest products-fuelwood, tanbark, charcoal, raw material for rayon manufacture, honey etc.
- Medicinal herbs
- Sites of fish ponds
- Sites for salt beds
- Sites for fish seed gathering
- Sites for local fisheries
- Sites for commercial and industrial establishments
- Sites for waste disposal

Ecological Values

- Source of food and shelter for commercially important fish and shellfish.
- Breeding and spawning grounds.
- Wild life habitat.
- Land building.
- Shoreline stabilization and protection-Good seed resources of prawns and fishes.
- Serves as a protective nursery ground for many species of fish, prawns, lobster, juveniles, etc.

Ecologically a mangrove ecosystem can be divisible into different zones. Zonation of different mangrove species from sea to land based on their silt and salinity tolerances is made into three zones.

1. The above-tide forest habitat
2. The intertidal-swamp habitat
3. The submerged-subtidal habitat

The Above Tide Forest Habitat

- This environment is inhabited by terrestrial species with trunk and leaf canopy.
- Birds, bats, lizards, tree snakes, snails, land crabs and mangrove crabs, spiders and insects are common residents with most diverse and abundant insects.
- The birds and bats are mostly insectivorous or piscivorous.
- The crabs are detritivorus or omnivores and may feed on marine food during low tide.
- Some areas are grazed by domestic animals (Cattle, goat, etc).

The Intertidal Swamp Habitat

- This environment offers a variety of different substrates and different microhabitates to support diverse community.
- Barnacles and oysters are conspicuous epifauna on the roots.
- Isopods bore into the Woody roots.
- Periwinkles (snails) abundant-crawling on sediment and roots.
- Polychaete warms associated with the root and tube building species on hard substrata.
- Burrowing fiddler crabs (Uca sp) are common on mud flats.
- Amphipods and some species of crabs feed on benthic algae, (macroalgae-*Padina gymnospara, Laurcncia papipllosn, Dictyota diclwtonm and Ulva lactitca,* Micro *algae-Enterotiiorpha Chiwluniorpha, Eladotnorplm and Ectocarpuse*)

Mud skippers (genus *Periophthalmus* fish burrow or crawl on the mud flats and can climb the mangroves. Many species of prawn and fish with tides.

The Subtidal Zone

- This zone has fine grained mud with a high organic content.
- Mangrove roots supports a rich epiflora and epifauna- Sponges, Tunicates, Anemones, Hydroids and Bryozoans.
- The turtle grass (Thalassia), sea grass (Zostera and Cymodocea) are the dominent benthic plants.
- The burrowing animals-crabs, shrimps warms are common.
- Plankton feeding fishes are common.
- The fish, as well as crabs, lobsters and shrimp form the basis for local fisheries.

Management

A national committee on wetlands mangroves and coral reefs constituted in 1993 by the Ministry of Environment and Forests, recommended certain guidelines for management of mangroves.

- Nationwide mapping of mangrove areas.
- Study on the growth rate and productivity of mangroves.
- Seasonal variation in climate and environment.
- Declaration of mangrove areas as "National parks".
- Afforestation of mangroves.
- Create awareness among public about the importance of mangroves.
- Identify Keystone species of trees, plants animals to each area.
- Establish base line data on water-soil characteristics, bio-diversity of flora and fauna of all mangrove systems. Establish number of trophic levels, energy transfer and ecosystem efficiency.

Mangrove associated Organisms

The mangrove ecosystem is associated with diverse type (more than 25 taxonomic groups) of algae, bacteria, fungi, flowering plants and animals (Table 23.14). Mangrove associated grasses and plants of salt marshes which are found to colonize mangrove areas and help to stabilize the soil and also to hold floating seeds and seedlings in a position for natural regeneration are shown below :

Table 23.14. Diversity of Different Taxonomic Groups Associated with Mangroves in India

Taxonomic Group	*No.*
Bacteria	20
Fungi	40
Algae	65
Bryophytes/Ferms	35
Lichens	2
Monocotyledons	73
Dicotyledons	110
Protozoa	30
Copepoda	50
Sponges	5
Ctenophara and Coelenterata	10
Bryozoa	20
Non-Polychaete worms	13
Nematodes	50
Polychaetes	11
Crustaceans	229
Insects and Arachnids	500
Molluscs	212
Echmoderms	1
Fishes	300

Reptiles	22
Amphibians	2
Birds	177
Mammals	36

Protersia coractata (Roxb), *Aeluropus lagopides* (Takaoka), *Llrochondra setulosa* (Hubbard), *Arthocnemuin indicurn* (Wild), *Salicornia brachiata* (Roxb), *Suaeda fruticosa, Suaeda maritima, Suaeda monica F.* and *Sesuvium portulacastrum L.*

Diatoms

01. *Amphora decussata*
02. *A. laevissima*
03. *A. Lineolata*
04. *A. Oiwlis*
05. *A. marina*
06. *Ahosatica*
07. *Achanthes brmipes*
08. *A. longipes*
09. *Atnphiphora gigantea*
10. *Biditlphia pitlchdla*
11. *B. sincnsis*
12. *Bacillaria paradoxa*
13. *Cnmpylodisciis binngnlatiis*
14. *Coscinodiscus centricns*
15. *Chadocews lorangiamif*
16. *C. janesianum*
17. *C. socialis*
18. *Cocconies littoralis*
19. *Climacosphcnia moniligera*
20. *Cymbella marina*
21. *Cetratulina sp*

22. *Dytilum brightwelli*
23. *Diploneis dydiina*
24. *D. Smithi*
25. *D. Splendida*
26. *D. subovalis*
27. *Epithemia niusculus*
28. *Fragillaria pinnata*
29. *Gyrosigma scalproides*
30. *Gyrosigma balticum*
31. *Grammatophora undulata*
32. *Hemidiscus sp*
33. *Isthtnia evervis*
34. *Licmophora flabellata*
35. *L. gracilis*
36. *Leptocylindricus donhts*
37. *Lithodesmium iindulatum*
38. *Mastogloia dolosa*
39. *M. exili*
40. *Melosira sulcata*
41. *Nitzschia longissima*
42. *N. seriata*
43. *N. Sigma*
44. *N. angustata*
45. *N. commutata*
46. *N. obtusa*
47. *N. plalea*
48. *A. punctata*
49. *N. vermicularis*
50. *Navicula hasta*
51. *N. lyra*

52. *JV. monilifera*
53. *N. permagna*
54. *A. granulata*
55. *N. digito radiata*
56. *Pleurosigma angulatum*
57. *P. directum*
58. *P. formusum*
59. *P. normanii*
60. *Pinnularia smithi*
61. *P. antbigue*
62. *P. interrupta*
63. *Plandtmella sol*
64. *Rhaphoneis discoides*
65. *Rhizosolemia cylindricum*
66. *R. setigera*
67. *Streptotheca jndica*
68. *Simrellu fluminensis*
69. *S. Striatua*
70. *Synedra formosa*
71. *S. pitlchella*
72. *S. ulna*
73. *Triceratium roberstianum*
74. *Trachyiteis aspera*
75. *Thallassiosira* sp
76. *Thallassiothrix* sp

Blue Green Algae

1. *Anabaena* sp
2. *Aplanocapsa* sp
3. *Coelosphaerium* sp
4. *Gleotricha* sp

5. *Merismopedia* sp
6. *Microcystis* sp
7. *Nostoc* sp
8. *Oscillatoria*
9. *Trichodesmium* sp
10. *Lyngbya*

Green Algae

1. *Cladophora* sp
2. *Netrium* sp
3. *Mongolia* sp
4. *Spirogyra*
5. *Microspora* sp

Dinoflagellates

1. *Certaium bergne*
2. *C. falcatum*
3. *C. macroceros*
4. *C. trichoes*
5. *Dinophysis miles*
6. *Peridinium*

Common Zooplankton

01. Calanoid Copepoda
02. Cyclopoid copepoda
03. Herpecticoid copepoda
04. Penilia sp
05. Evadne sp
06. Conchecia sp
07. Anthrornedusae
08. Leptornedusae
09. Pleurobranchia sp
10. Sagitta sp

11. Foraminiferids
12. Tintinnids
13. Rotifers
14. Mysids
15. Oikopleura sp
16. Protozoa
17. Nauplius-copepod
18. Nauplius-prawn
19. Nauplius-barnacles
20. Protozoea
21. Mysis larvae
22. Zoea
23. Megalopa
24. Bivalve larvae
25. Gabiropfxi larvae
26. Polychaete larve
27. Fish larvae
28. Fish eggs
29. Lucifer

Common Macrobcnthos

1. *Dcndroncreis aestuarina*
2. *D. Arborifera*
3. *Lumbriconereis polydesma*
4. *L. notocirrata*
5. *Lycastis indica*
6. *Nereis-glandicincta*
7. *N. chilkensis*
8. *Nephthys polybranchia*
9. *N. oligobranchia*

Molluscans

1. *Merdrix casta*
2. *Cerithedea fluviatilis*
3. *Telescopium telescopium*
4. *T. fuscum*
5. *Bivalve larvae*
6. *Gastropod larvae*

Crustacea

1. Uca sp
2. Mysids
3. Apseudus sp
4. Limnoria sp
5. Amphelisca sp
6. Corophium sp
7. Alima larvae
8. Mysis larvae
9. Megalopa
10. Post larvae

Common Shell Fishes and Fin Fishes

01. Mugil *cephalus*
02. Liza *macrolepis*
03. Liza *tade*
04. Liza parsifl
05. *Valamugil seheli*
06. *V. Spciglerri*
07. *Genes filamcntosus*
08. *Sillago sihaina*
09. *Eutroplus suratensis*
10. *Chanos chanof*
11. *Mcgalopsis cardyla*

12. *Ambasis sp*
13. *Tylosurus* sp
14. *Nematolopa nasus*
15. *Pomadasys* sp
16. *Thrissocles mytax*
17. *Anadontostoma chacunda*
18. *Penaeus indicus*
19. *P. merguiensis*
20. *Metapenaeus monoceros*
21. *M. dobsoni*

Fish and Shrimp Seeds

01. *Mugil cephalus*
02. Liza niacrolepis
03. Liza tade
04. Liza parsw
05. *Sillago sihama*
06. Gerrs setifer
07. Gerus *filamentosus*
08. Etroplus *aurus*
09. *Chanos chanos*
10. *Petweus indicus*
11. *P. merguinensis*
12. *Metapenaeus monoceros*
13. *Metapcnans dobsoni*

Pic. 23.1. Tribal women near the site of the refinery.

Pic. 23.2. Seasonal workers from Karnataka in a paddy field near Fort Aguada. They hail from Zinid taluk in Bijapur district and were forced to migrate by drought conditions.

8

PLANT PROTECTION INTERNATIONAL CONVENTION

Preamble

The Contracting Governments, recognizing the usefulness of international cooperation in controlling pests and diseases of plants and plant products and in preventing their introduction and spread across national boundaries, and desiring to ensure close coordination of measures directed to these ends, have agreed as follows :

Article I

Purpose and Responsibility

1. With the purpose of securing common and effective action to prevent the introduction and spread of pests and diseases of plants and plant products and to promote measures for their control, the contracting Governments undertake to adopt the legislative, technical and administrative measures specified in this Convention and in supplementary agreements pursuant to Article III.
2. Each contracting Government shall assume responsibility for the fulfilment within its territories of all requirements under this Convention.

Article II

Scope

1. For the purposes of this Convention the term "plant" shall comprise living plants and parts thereof, including seeds in so far as the supervision of their important under Article VI of the Convention or the issue of phytosanitary certificates in respect of them under Articles IV(1), (a) (iv) and V of this Convention may be deemed necessary by contracting Governments; and the term "plant products" shall comprise unmanufactured and milled material of plant origin, including seeds in so far as they are not included in the term "plants".
2. The provisions of this Convention may be deemed by contracting Governments to extend to storage places, containers, conveyances, packing material and accompanying media if all sorts including soil involved in the international transportation of plants and plant products.
3. This Convention shall have particular reference to pests and diseases of importance to international trade.

Article III

Suplementary Agreements

1. Supplementary agreements applicable to specific regions, to specific pests and diseases, to specific plants and plant products, to specific methods of international transportation of plants and plant products, or otherwise supplementing the provisions of this Convention, may be proposed by the Food and Agriculture Organization of the United Nations (hereinafter referred to as "FAO") on the recommendation of contracting Government or on its own initiative, to meet special problems of plant protection which need particular attention or action.

2. Any such supplementary agreements shall come into force for each contracting Government after acceptance in accordance with the provisions of the FAO Constitution and Rules of Procedure.

Article IV

National Organization for Plant Protection

1. Each contracting Government shall make provision, as soon as possible and to the best of its ability, for

(a) an official plant protection organization, with the following main functions:

(i) the inspection of growing plant, of areas under cultivation (including fields, plantations, nurseries, gardens and green houses), and of plants and plant products in storage and in transportation particularly with the object of reporting the existence, outbreak and spread of plant diseases and pests and of controlling those pests and diseases;

(ii) the inspection of consignments of plants and plant products moving in international traffic, and, as far as practicable, the inspection of consignments of other articles or commodities moving in international traffic under conditions where they may act incidentally as carriers of pests and diseases of plants and plant products, and the inspection and supervision of storage and transportation facilities of all kinds involved in international traffic whether of plants and plant products or of other commodities, particularly with the object of preventing the dissemination across national boundaries of pests and diseases of plants and plant products;

(iii) the disinfestation or disinfection of consignments of plants and plant products moving in international traffic, and their containers, storage places, or transportation facilities of all kinds employed;

(iv) the issue of certificates relating to phytosanitary conditions and origin of consignments of plants and plant products (hereinafter referred to as "phytosanitary certificates");

(b) the distribution of information within the country regarding the pests and diseases of plants and plant products and the means of their prevention and control;

(c) research and investigation in the field of plant protection.

2. Each contracting Government shall submit a description of the scope of its national organization for plant protection and of changes in such organization to the Director-General of FAO, who shall circulate such information to all contracting Governments.

Article V

Phytosanitary Certificates

1. Each contracting Government shall make arrangements for the issue of phytosanitary certificates to accord with the plant protection regulations of other contracting Governments and in conformity with the following provisions:

(a) Inspection shall be carried out and certificates issued only by or under the authority of technically qualified and duly authorized officers and in such circumstances and with such knowledge and information available to those officers that the authorities of importing countries may accept such certificates with confidence as dependable documents.

(b) Each certificate covering materials intended for planting or propagation shall be as worded in the Annex to this Convention and shall include such additional declarations as may be required by the importing country. The model certificate may also be used for other plants or plant products where appropriate and not inconsistent with the requirements of the importing country.

(c) The certificates shall bear no alterations or erasures.

2. Each contracting Government undertakes not to require consignment of plants intended for planting or propagation imported into this territories to be accompanied

by phytosanitary certificates inconsistent with the model set out in the Annex to this Convention.

Article VI

Requirements in Relation to Imports

1. With the aim of preventing the introduction of diseases and pests of plants into their territories, contracting Governments shall have full authority to regulate the entry of plants and plant products, and to this end, may:

(a) prescribe restrictions or requirements concerning the importation of plants or plant products;

(b) prohibit the importation of particular plants or plant products, or of particular consignments of plants or plant products;

(c) inspect or detain particular consignments of plants or plant products;

(d) treat, destroy or refuse entry to particular consignments of plants or plant products, or require such consignments to be treated or destroyed.

2. In order to minimize interference with international trade, each contracting Government undertakes to carry out the provisions referred to in paragraph 1 of this Article in conformity with the following:

(a) Contracting Governments shall not, under this plant protection legislation, take any of the measures specified in paragraph 1 of this Article unless such measures are made necessary by phytosanitary considerations.

(b) If a contracting Governments prescribes any restrictions or requirements concerning the importation of plants and plant products into its territories, it shall publish the restrictions or requirements and communicate them immediately to the plant protection services of other contracting Governments and to FAO.

(c) If a contracting Government prohibits, under the provisions of its plant protection legislation, the importation of any plant or plant products, it shall publish its decision with reasons and shall immediately inform the plant protection services of other contracting Governments and FAO.

(d) If a contracting Government requires consignment of particular products to be imported only through specified points of entry, such points shall be so selected as not unnecessarily to impede international commerce. The contracting Government shall publish a list of such points of entry and communicate it to the plant protection services of other contracting Governments and to FAO, such restrictions on points of entry shall not be made unless the plants or plant products concerned are required to be accompanied by phytosanitary certificates or to be submitted to inspection or treatment.

(e) Any inspection by the plant protection service of a contracting Government of consignments of plants offered for importation shall take place as promptly as possible with due regard to the perishability of the plants concerned. If any consignment is found not to conform to the requirements of the plant protection legislation of the importing country, the plant protection service of the exporting country shall be informed. If the consignment is destroyed, in whole or in part, and official report shall be forwarded immediately to the plant protection service of the exporting country.

(f) Contracting Governments shall make provisions which, without endangering their own plant production, will reduce to a minimum the number of cases in which a phytosanitary certificate is required on the entry of plants or plant products not intended for planting, such as cereals, fruits, vegetables and cut flowers.

(g) Contracting Governments may make provision for the importation for purposes of scientific research of plants and plant products and of specimens of plant pests and disease-causing organisms under conditions affording ample precaution against the risk of spreading plant diseases and pests.

3. The measures specified in this Article shall not be applied to goods in transit throughout the territories of contracting Governments unless such measures are necessary for the protection of their own plants.

Article VII

International Co-operation

The contracting Governments shall co-operate with one another to the fullest practicable extent in achieving the aims of this Convention, in particular as follows:

(a)Each contracting Government agrees to co-operate with FAO in the establishment of a world reporting service on plant diseases and pests, making full use of the facilities and services of existing organizations for this purpose, and, when this is established, to furnish to FAO periodically the following information.

(i) reports on the occurrence, outbreak and spread of economically important pests and diseases of plants and plant products which may be of immediate or potential danger;

(ii) information on means found to be effective in controlling the pests and diseases of plants and plant products.

(b) Each contracting Government shall, as far as is practicable participate in any special campaigns for combating particular destructive pests or diseases which may seriously threaten crop production and need international action to meet the emergencies.

Article VIII

Regional Plant Protection Organization

1. The contracting Governments undertake to co-operate with one another in establishing regional plant protection organizations in appropriate areas.

The regional plant protection organizations shall function as the co-ordinating bodies in the areas covered and shall participate in various activities to achieve the objectives of this Convention.

Article IX

Settlement of Disputes

1. If there is any dispute regarding the interpretation or application of this Convention, or if a contracting Government considers that any action by another contracting Government is in conflict with the obligations of the latter under Articles V and VI of this Convention, especially regarding the basis of prohibiting or restricting the imports of plants or plant products coming from its territories, the Government or Governments concerned may request the Director-General of FAO to appoint a committee to consider the question in dispute.
2. The Director-General of FAO shall thereupon, after consultation with the Governments concerned, appoint a committee of experts which shall include representatives of those Governments. This committee shall consider the question in dispute, taking into account all documents and other forms of evidence submitted by the Governments concerned. This committee shall submit a report to the Director-General of FAO who shall transmit it to the Governments concerned, and to other contracting Governments.
3. The contracting Governments agree that the recommendations of such a committee while not binding in character, will become the basis for renewed consideration by the Governments concerned of the matter out of which the disagreement arose.
4. The Governments concerned shall share equally the expenses of the experts.

Article X

Substitution of Prior Agreements

This Convention shall terminate and replace, between contracting Governments, the International Convention respecting measures to be taken against the Phylloxera vastarix of 3 November, 1881, the additional Convention signed at Berne of 15 April, 1889 and the International Convention for the Protection of Plants signed at Rome on 16 April, 1929.

Article XI

Territorial Application

1. Any Government may at the time of ratification or adherence or at any time thereafter communicate to the Director-General of FAO a declaration that this Convention shall extend to all or any of the territories for the international relations of which it is responsible, and this Convention shall be applicable to all territories specified in the declaration as from the thirtieth day after the receipt of the declaration by the Director-General.
2. Any Government which has communicated to the Director-General of FAO a declaration in accordance with paragraph 1 of this Article may at any time communicate a further declaration modifying the scope of any former declaration or determinating the application of the provisions of the present Convention in respect of any territory. Such modification or termination shall take effect as from the thirtieth day after the receipt of the declaration by the Director-General.
3. The Director-General of FAO shall inform all signatory and adhering Governments of any declaration received under this Article.

Article XII

Ratification and Adherence

1. This Convention shall be open for signature by all Governments until 1 May 1952 and shall be ratified

at the earliest possible date. The instruments of ratification shall be deposited with the Director-General of FAO, who shall give notice of the elate of deposit to each of the signatory Governments.

2. As soon as this Convention has come into force in accordance with Article XIV, it shall be opened for adherence by non-signatory Governments. Adherence shall be effected by the deposit of an instrument of adherence with the Director-General of FAO, who shall notify all signatory and adhering Governments.

Article XIII

Amendment

1. Any proposal by a contracting Government for the amendment of this Convention shall be communicated to the Director-General of FAO.

2. Any proposal amendment of this Convention received by the Director-General of FAO from a contracting Government shall be presented to a regular or special session of the Conference of FAO for approval and, if the amendment involves important technical changes or imposes additional obligations on the contracting Governments, it shall be considered by an advisory committee of specialists convened by FAO prior to the Conference.

3. Notice of any proposed amendment of this Convention shall be transmitted to the contracting Governments by the Director-General of FAO not later than the time when the agenda of the session of the Conference at which the matter is to be considered is dispatched.

4. Any such proposed amendment of this Convention shall require the approval of the Conference of FAO and shall come into force as from the thirtieth day after acceptance by two-thirds of the contracting Governments.

5. The instruments of acceptable of amendments involving new obligations shall be deposited with the Director-General of FAO, who shall inform all contracting Governments of the receipt of acceptances and the entry into force of amendments.

Article XV

Denunciation

1. Any contracting Government may at any time give notice of denunciation of this Convention by notification addressed to the Director-General of FAO. The Director-General shall at once inform all signatory and adhering Governments.
2. Denunciation shall take effect one year from the date of receipt of the notification by the Director-General of FAO.

9

BIODIVERSITY IN FARMING

Agricultural progress seems even more paradoxical when we consider biodiversity use in agriculture. Several thousand plant and animal species are known to have served as food in human history. The displacement of traditional crop varieties by newly-developed highyielding varieties has been a major factor in this development. FAO estimates that in the next 10 years 90% of the existing genetic diversity within the main crops is at serious risk of disappearing. The vast majority of the literature dealing with biodiversity use in agriculture is characterized by a symptomatic approach, and centers on providing data that prove the continuous reduction of both inter and intra-species biodiversity used in food production on its associated risks for the sustainability of our food supply and on alternative farming techniques to halt the erosion of biodiversity use in agriculture.

A discussion of the driving forces that have generated the reduction of biodiversity use in agriculture is generally missing. Put another way, the detection of symptoms is directly coupled to a prescription of remedies without any serious attempts to diagnose that disease. Perhaps, this is because a thorough diagnosis would necessarily raise some politically unpleasant issues. The decreasing energy output/input ratio in agriculture means that the food security of humankind is increasingly dependent on rapidly dwindling oil reserves and on an increasing environmental impact per unit of crop output. The dramatic reduction of biodiversity worldwide spells serious problem for the maintenance of ecological, equilibria at the planet level. That is the path of technological in food production chosen right now by humankind is not sustainable. Not only are these unpopular issues to discuss, but it is becoming evident that most of the undesirable side effects leading to unsustainability are difficult, if not impossible, to incorporate in conventional sets of measures and indicators used to assess risks, costs and benefits of technological change. In spite of these difficulties, it is crucial to understand the driving forces that are shaping current trends of technological development in otherwise, we seriously risk caveats in discussions of future agricultural policies. It is common knowledge that traditional techniques achieved high output/ input energy ratios and used much more biodiversity than modern high input agriculture.

Individual scientific disciplines that study change in socio-economic systems tend to focus on the effect of one factor at the time. Although, this approach provides a high resolution for the factor in question, it does not shed much light on the bigger picture. In this respect, the agricultural sciences are no exception.

Ecologists tend to focus on the on-site and off-site effects of agriculture on the environment without considering agriculture as a livelihood. Economists are concerned with the economic return of the labour and capital resources invested in the farming process, and tend to ignore the ecological impacts. Agronomists occupy themselves basically with the physical performance of the production system. Rural sociologists and anthropologists are mainly interested in the effects of technological changes on the lives of individual farmers and the identity of rural communities. This specialization or reductionists of established scientific disciplines is at times necessary, but often creates confusion for decision-makers that have to decide about the future directions of technical development in agriculture.

All of the above-mentioned aspects of agriculture must be addressed since each one is fundamental for the sustainability of agriculture. The problem lies in the fact that a detailed discussion of any one of these aspects becomes useless if not put into relation to the functioning of the system as for this reason, it is of utmost importance to put the different disciplines in perspective within a common frame of analysis.

The existence of different perspectives for assessment of sustainability of agricultural development has been pointed out by many workers. In spite of minor differences in definitions and emphasis among the different authors, there is substantial agreement that at least three fundamental perspectives should be considered when attempting to define sustainable agriculture:

- The ecological view: agricultural techniques must be environmentally sound.
- The economic view: agricultural techniques must be economically viable.
- The social view: agricultural techniques must be acceptable to farmers and society given culture, ethics and religon.

Assessing only agronomic performance or ecological compatibility of various agricultural techniques is not sufficient to evaluate their sustainability. We need a parallel check on economic viability and social acceptability. The farming system as a sub-system of society: constraints imposed by population density, income level and labour time allocation in society. In order to verify whether certain agricultural production techniques are convenient for a given society, we should first analyse how the organization and technological development of society influence the way in which food is produced in its agricultural sector. Two main characteristics of society that I will examine in this paper are:

1. Population density, or demographic pressure. One of the most obvious aspects of demographic pressure that directly influences the agricultural sector is the amount of arable land per capita available for agricultural production.

2. The level of technological development of a society, or socio-economic pressure. This can be conveniently quantified by the following two indicators:

- The average per capita income or average per capita commercial energy consumption ill society.;
- The fraction of total human time available that is allocated to work in the primary sectors of the economy. Of the primary economic sectors, agriculture is one of the most affected in terms of contraction of human time allocation. A cross-section analysis over a sample of 60 countries for the years 1990-1991 shows that the fraction of the labour force engaged in agriculture is inversely correlated with the GDP per capita. Countries with GDP per capita of more than US $ 10000 per year have less than 7% of the working force in agriculture, and the richest countries even as little as 2%.

The demographic pressure or population density of society is simply defined as the ratio. It is well known that population density has been an important factor in shaping the development of agricultural production techniques. Apart from scientific analyses, common sense also suggests that a high demographic pressure in society will favour farming techniques that yield a high food production per unit area.

The effect of demographic pressure on agricultural production can be expressed by the indicator AP_{DP} (=agricultural productivity imposed by demographic pressure), which is defined as the productivity of land (yield of food energy per hectare) that would be needed to obtain a situation of complete food self-sufficiency in society (matching the aggregate food requirement of society). Thus, defined numerical assessments of AP_{DP} require knowledge of the aggregate requirement of food in society (considering the food system under analysis as closed), which is in turn determined by the population size of society, the food consumption pattern, and post-harvest losses. The amount of land available for food production, which depends on availability of arable land, characteristics of arable land, and alternative land-uses (dependent on population size and level of technical development).

The effect of socio-economic pressure on agriculture can be expressed by the indicator APSEP agricultural productivity imposed by socio-economic pressure, which is defined as the productivity of labour in agriculture (yield of food energy per hour of labour in the agricultural sector) that would be needed to obtain food self-sufficiency for society.

Considering society solid system as closed, APSEP is obtained by dividing the yearly aggregate requirement of food in society by the labour time available in a year for generating the food supply depends on the size of the labour force, the unemployment rate, tile fraction of the labour force absorbed by the non-agricultural sectors, and the average work load. The US economy, with only 2% of its workers engaged in agriculture, has an APSEP of 270 MJ of food energy per hour of agricultural labour. This value is well out of the range of possible productivity of farmers in developing countries (less than 10 W) and is not even reached by farmers in the European Union. The requirement for food in society is proportional to population size, and therefore to the total human time available to society, whereas internal food production is proportional to the labour time available in the agricultural sector. Hence, in order to achieve and maintain food self-sufficiency, any reduction in the fraction of total labour time allocated to the sector must be compensated by an increase in the amount of food produced per hour of labour in the agricultural sector. According to this rationale, a socio-economic structure that allocates only a small fraction of its labour force to the agricultural sector will select farming techniques that result in a high food production per unit of labour in the agricultural sector.

The indicators AP_{DP} and AP_{SEP} represent the minimum density of biophysical throughput, per hectare and per hour, respectively, that would be needed on average in the agricultural sector to achieve a state of food self-sufficiency in a defined society. Hence, they are based on characteristics of the socio-economic system as a whole, and their value can be calculated without any knowledge of what is going on at the level of the fanning system in the society. How, then, can these indicators be related to the actual technological changes occurring at the farm level? Are the values of the indicators AP_{DP} and AP_{DP} that characterize the pressure that society exerts on its fanning systems, effectively related to the techniques of food production realized at the farm level? To answer these questions, it is useful to compare for several types of societies, the values of AP_{DP} and AP_{SEP} with actual values achieved in agricultural production in these societies. For instance, the minimum productivity of land imposed by demographic characteristics (AP_{DP} can be compared with the actual agricultural productivity per hectare (APha) for the country as a whole (average), or at the regional level, or for individual farms. The actual productivity of land or yield of food energy per hectare achieved by a country, or farming system is obtained by dividing the aggregate production of food within the defined territory by the amount of land used in the production process. Similarly, expected minimum value of labour productivity in agriculture according to socio-economic characteristics can be compared with the actual agricultural productivity (throughput of food energy) achieved per hour of labour in agriculture for the country as a whole (average) or for selected farming systems. Such a comparative analysis, at the national level has been carried out for a large sample of countries, to see whether the demographic and socio-economic pressure are indeed related to the way in which food is actually produced in different socio-economic contexts.

The analysis shows that even if the two indicators do not indicate mandatory threshold values, they should in any case be considered attraction values, toward which the agricultural system feuds to move in practice. If a country has a high demographic pressure it is also likely to have a large production of food energy per hectare in the agricultural sector.

Countries not achieving self-sufficiency must rely on imports or food aid to guarantee food security for their population. The analysis described above indicates the existence of a link between the variables AP_{DP} and AP_{SEP} on the other hand, that are defined at the hierarchical level of society as a whole, and the variables AP_{DP} and AP_{HOUR} on the other hand, that are defined and measured at the hierarchical level of farming systems. This link suggests that it is possible to describe technical changes in food production on both levels on the same geometric plane. This plane is defined by two axes, the density of agricultural throughput per hour of labour and the density of agricultural throughput per hectare of land.

Considering the process of agricultural production, changes in APha and AP_{hour} reflect changes in technical coefficients of technologies adopted at the farming system level. On the higher level, society as a whole, changes in AP_{DP} and AP_{hour} reflect alterations in the characteristics of the socio-economic system to which the agricultural sector belongs. Being part of both levels, agriculture must operate in such a way that the combination of values of AP_{DP} and AP_{hour} is compatible with the constraints imposed by society, that is the combination of the values of AP_{DP} and AP_{SEP}. Clearly, farming systems of subsistence societies occupy in area of the plane that is quite distant from that occupied by farming systems of industrialized countries. In general, farming systems tend to have an actual agricultural performance that is closely related to the set of socio-economic constraints corresponding to the socio-economic context in which they are operating. Describing changes in agricultural production on the plane formed by the agricultural throughput per hour labour and the throughput per hectare of land makes it possible to define trajectories of changes for both socio-economic variables and technological coefficients in agriculture, which are expected to accompany trends in technological development. For instance, current trends in economic development suggest an economic growth of developing countries that is characterized by increases in both population size and per capita consumption. This will be reflected in a parallel increase in AP_{DP} and AP_{SEP}. Sooner or later, these changes in demographic and socio-economic pressure (AP_{DP} and AP_{SEP}) will be reflected into altered technical coefficients of farming techniques. That is, technological development in agriculture in this situation tends to evolve toward higher values of AP_{DP} and AP_{hour}.

General trends in the evolution of food production techniques for different types of socio-economic systems can be represented on the two-dimensional plane. Four main types of socio-economic systems, having different combinations of demographic and socio-economic pressure.

- *Societies with low demographic and socio-economic pressure:* The population is growing faster than the GNP per capita, which means that AP_{DP} will grow faster than AP_{SEP}. Hence, they will move toward a situation typical of PR China.
- *Societies with low demographic and high socio-economic pressure:* Economic development is expected to be maintained and population growth will be relatively slow but steady. On the plane, this means a slow movement toward higher values of AP_{DP}.
- *Societies with high demographic and low socio-economic pressure:* They tend to show economic growth and maintain if not expand their already huge population size. At national level, an increasing GNP per capita will result in an accelerated absorption of the labour force currently engaged in agriculture by other sectors of the economy. This will inevitably require a major increase in agricultural labour productivity to maintain security. Hence, a movement toward the West European conditions of agricultural production is to be expected.
- *Societies with high demographic and socio-economic pressure:* These societies have no alternative but to

try to maintain a high material standard of living and keep population growth to a minimum. This means a more or less stable and high level of APSRJ and a very slowly increasing value of AP_{DP}.

For these societies, trying to reduce the environmental impact of their food production becomes a major factor. Note that food import from the international market, a must for countries where $AP_{DP} > AP_{ha}$ and/or $AP_{SEP} > AP_{hour}$ is based on the existence of surpluses produced by countries where the relation between these parameters is inverse. At present, the USA, Canada, Australia and Argentina combined 80% of the net export of cereals in the world market.

However, at their present rate of population growth and because of an increasing concern for the environment this surplus may be seriously eroded in the near future, for instance, the USA is expected to double its population in 60 years. Food import problems will be accelerated if a general concern for the environment will lead to a less intensive agriculture all over the world. At world level, demographic and socio-economic pressure are certainly expected to increase. It is often overlooked that, at world level, there is no option to import food from elsewhere. When increases in demographic and socio-economic pressure are not matched by an adequate increase in productivity of land and labour, food imports of the rich will be based on starvation of the poor. Worldwide, arable land per capita is already less than 0.27 ha per capita at present, and is expected to continue to shrink because of population growth.

In addition, arable land is being lost. For instance, during the past 40 years nearly one-third of the world's cropland has been abandoned because of soil erosion and other types of degradation. Most of the added land that replaces this loss has come from marginal land made available by deforestation.

The increasing dependence of the food security of poor densely populated countries on oil may become problematic in the near future. In the last few years, PR China and other Asian countries have been among the countries with the fastest growing economics in terms of GNP. At world level, fuelling the economic development of densely populated countries represents a major drain on natural resources, notably fresh water and fossil energy stocks. Not only do we risk that a dramatic surge in oil demand will augment oil prices, but we may soon face the bottom of the oil barrels.

The goal of achieving sustainability of food production in the future is closely related to the challenge of mediating between the socio-economic constraints 'imposed' by society, that push for more concentrated and faster agricultural throughputs, and ecological constraints 'imposed' by natural processes that push in the opposite direction.

Natural processes provide the services on which all agricultural production relies, such as water and nutrient supply, oil fertility, pollination, stabilization of atmospheric composition, and regulation of biogeochemical cycles. Respecting ecological constraints means keeping the flows of biomass produced in the agro-ecosystem by human management within a range of values that is compatible with regulation mechanisms typical of natural ecosystems.

A large discrepancy between the actual density of agricultural throughput and the densities imposed by either socio-economic or ecological constraints indicates unsustainability of agricultural production. Agriculture is ecologically unsustainable when the actual density of agricultural throughput exceeds the critical environmental loading, and socio-economically unsustainable when the actual density of agricultural throughput is too low to provide food security and/or economic viability for farmers.

Agriculture can be defined as a human activity that exploits natural processes and natural resources in order to obtain food and other products considered useful by society. The verb 'exploits' suggests that we deal with an alternation. Indeed, within a defined area humans alter the natural distribution of animal and plant populations in order to selectively increase the density of certain flows of biomass that they consider more useful for the socio-economic system.

Adopting this definition, we implicitly assume that agriculture has to do with manipulation of biodiversity. Following the schema of ecosystem structure proposed by Odum, a natural ecosystem can be seen as a network of matter and energy flows in which nutrients are mainly recycled within the system and solar energy is used to sustain this matter cycling.

The amount of solar energy used for self-organization by the ecosystem is proportional to the speed of its matter cycles (in terrestrial ecosystems this also has to do with availability and circulation of water). Agriculture then involves a boosting of only those matter and energy flows in the network that humans consider beneficial, and an elimination or reduction of the flows that they consider detrimental to their purposes. In doing so, humans amplify some of the existing genetic information (that is contained in populations representing 'useful' biomass), and reduce or eliminate some other genetic information. In other words, changes in the hierarchical structure of matter and energy flows in the ecosystem must be coupled to changes in the hierarchical structure of genetic information clustered in species and distributed over tax in the geographical space altered for agricultural purposes. The hierarchical structure of genetic information can be stabilized only by a quasi-steady-state in nutrient flows, which in turn depends on the ability to use available flows of free energy.

The resonance between coded information in DNA and the metabolism of living systems implies that the reverse must also be true. The particular hierarchical structure of matter and energy flows is determined by the rate of input/output of individual components of the network and their spatial and quantitative distribution; that is, on a large scale, it depends on the hierarchical organization of genetic information. Profound changes in the densities of harvested biomass can have dramatic consequences for the ecosystem structure. The elimination of ecosystem diversity can affect the stability of the biosphere when the scale of human alteration is increased to reach global relevance.

For example, in order to significantly increase yields of biomass (the amount taken away from the agro-ecosystem per unit of area and time), it is necessary to fight competing species (pests) and to bypass physiological rates of nutrient cycling in the ecosystem (fertilization) when these become limiting. This calls for selection of crop species and varieties that perform well tinder human managed conditions.

The interaction of the agricultural sector with both ecosystem and society can be studied by following inputs and outputs of the four main steps of the food system cycle:

- Food production, where food is defined as energy and matter compatible with human metabolism.
- Making the food accessible to consumers, where accessible food is defined as meals ready to be consumed according to defined food consumption patterns.
- Consuming food and generating wastes, where wastes are defined as energy and matter no longer compatible with human needs.
- Recycling wastes to agricultural inputs, where agricultural inputs are defined as energy and matter compatible with the agricultural process.

All steps are interconnected, and hence the definition of a flow as an input, an available resource, an accessible resource, or waste, is arbitrary. First of all, the definition depends on the point of view from which the system is analysed. For example, the introduction of trees leads to increased evapotranspiration, which may be negative in terms of less accessible water in the soil, but positive in terms of more available water in the form of rain clouds. Second, the definition depends on the compatibility of the throughput density with the processes regulating the particular step in question.

For example, human excrement of a small Chinese village are valuable input for agriculture, but the sewage of a big city is a major pollution problem. Whether or not the speed of a throughput at any particular step is compatible with the system as a whole depends on the internal organization of the system. Nobody expects science to make it possible to feed a person 30000 kcal of food per day, about 10 times the normal amount, while keeping that person alive for a long period of time. Why then do many believe it to be possible to increase the productivity of agro-ecosystems by several times without generating any, negative side-effects on agro-ecosystem health?

Indeed, the increasing intensity of agricultural throughputs over increasing areas of our planet should be considered a major threat to the conservation of biodiversity. The more the amount of agricultural biomass harvested from a defined population (considering a single species at the time) in an altered ecosystem differs in density from the biomass turnover of a similar species in naturally occurring ecosystems, the larger can be expected to be the environmental impact on the agro-ecosystem and the energetic cost of the agricultural production.

Extremely high densities of agricultural throughput (per hectare and/or per hour) necessarily require production techniques that ignore the functional mechanism of natural ecosystems, that is the nutrient cycles powered by a linear flow of solar energy. Clearly, individual species, taken one at the time, cannot generate cycles in matter, since they perform only a defined 'job' (occupy a certain niche) in the ecosystem.

Technical progress sooner or later implies a switch from low-input agriculture based on nutrient cycling within the agro-ecosystem to high input agriculture in which the throughput density of harvested biomass is directly controlled and maintained at elevated levels through reliance on external inputs that provide linear flows of both nutrients and energy. In low-input agriculture, where the harvested flow of biomass remains within the range of natural turnover values, the activity of production interferes only to a limited extent with the system of controls regulating matter and energy flows in the ecosystem. This form of agriculture requires that at least several distinct species are used in the process of production so as to maintain the internal cycling of natural inputs as a pillar of the agricultural production process.

For humans, this implies a poor control over the flow of produced biomass in the agro-ecosystem because of the low productivity per hectare when assessing the yield of a particular crop at the time and the variability in productivity in case of perturbations. Especially serious is the problem of facing dramatic increases of demographic pressure. This type of farming cannot work properly with levels of throughput per hectare too high.

On the positive side, direct and indirect biophysical costs of production are low. Few technical inputs are required per unit of output produced and, when population pressure is not too high, the environmental impact of this form of agricultural production is modest. The contrary is true for high-input agriculture. In this case, the harvested flow of biomass is well out of the range of values that is compatible with regulation processes typical of natural ecosystems (when considering yields of a single cultivated species at the time). In high-input agriculture, human management is based on an eradication of the natural structure of controls in the ecosystem.

In fact, where several tonnes of grain have to be produced per hectare and hundreds of kilograms of grain per hour of agricultural labour, natural rates of nutrient cycling and a natural structure of biological communities are unacceptable. In high-input agriculture, not even the genetic material used for agricultural production is related to the original ecosystem in which production takes place.

In this way, humans keep the flow of produced biomass tightly under control and adjust yields to match increasing demographic and socio-economic pressure (e.g. the green revolution). However, this control is paid for with a high energetic cost of food production, because humans must continuously provide artificial regulation in the agro-ecosystem in the form of inputs and defense against undesired, competitive species.

The environmental impact of high-input food production is necessarily large, and involves a dramatic reduction of biodiversity in the altered area, destruction of the entire set of mechanisms that regulated ecosystem functioning before alteration, and several negative side-effects such as on-site and off-site pesticide and fertilizer pollution, soil erosion, and salinization.

How do changes in technical coefficients of farming techniques, aimed at an increase of both land and labour productivity, affect biodiversity use in agriculture. What is the nature of trade-off between biodiversity use in agriculture on the one hand and productivity of land and/or labour on the other hand? At the level of the farming system the parameters land productivity and labour productivity are interdependent. The existence of this link is illustrated by the possibility to express the parameter labour productivity as, the product of two terms, one of which is the productivity of the land itself:

$$\text{labour productivity (kg hr}^{-1}) = \text{land productivity (kg ha}^{-1}) \times \text{land managed per hour (ha hr}^{-1})$$

Hence labour productivity can be augmented by either increasing the land area managed per hour of labour, increasing the land productivity or a combination of these two solutions. Usually, technical progress induces a combination of the two solutions which are mutually reinforcing.

The land managed per hour of labour can be augmented, while maintaining the same yield per hectare, by a substitution of machine power for human power. This solution pushes for monocultures because a monocultures facilitates mechanization by, synchronizing and concentrating in time and space the requirement of inputs and the harvest of outputs. With this solution, labour productivity can be raised by using more land per farmer, while avoiding all increase in energy costs per hectare in the form of land-saving inputs such as fertilizers and irrigation. This solution is typical of farming systems in Canada and Australia, where the demographic pressure is low.

Since capital substitution for labour (human power) must have a positive economic return to pay back its own investment, economic feasibility is a crucial factor for this solution. There are two obvious ways to increase farm revenues in this situation. First, by trying to get the most out of the local conditions by focusing on the crop(s) that show(s) the best comparative advantage in the area.

Second, by trying to get the most out of the capital investment by expanding the area of the farm, carefully concentrating cultivation in favourable spots and avoiding marginal lands. To facilitate mechanization, farmers generally tend to increase the dimension of their fields, by eliminating patchworks and edges, despite the fact that those landscape patterns and elements are known to be important in the maintenance of beneficial insect species for pest control and conservation of biodiversity in general.

Thus, the use of machine power to increase labour productivity will drive technological development in agriculture further toward the exploitation of comparative advantages and economies of scale (e.g. enlargement of farms). The building of specialized technical infrastructures in the area will reinforce the trend of product specialization and generate a mode-locking effect.

While increasing yields per hectare without enlarging the land area managed per unit of labour can be a (temporary) solution to match an increasing demographic pressure, it has limited power to meet the demand for a higher labour productivity. The problems with this solution are evident in PR China, where massive fertilization and irrigation are practiced to augment yields coupled to labour-intensive farming techniques. Although in this way agriculture will feed a densely-populated country, it will provide little support for its socio-economic development.

The equation of labour productivity indicates that both an increase in land productivity and an increase in land managed per hour of labour can have a positive marginal return in terms of an increased labour productivity and economic return. Therefore, the particular choice of technical inputs in agricultural production-machinery to boost the land managed per unit of labour or fertilizers and irrigation to boost land productivity-will depend on the expected marginal return of capital investments in these two factors, Sparsely-population developed countries, such as Canada, the USA and Australia tend to invest more in machinery, even though a certain investment in landsaving inputs is made when the marginal return is high.

Densely-populated developing countries, such as PR China, Egypt and Bangladesh, tend to predominantly invest in fertilizers and irrigation, and invest only little in machinery. Densely-populated wealthy countries, such as Japan, Italy and the Netherlands, are in a peculiar situation. In these countries, farming can become a 'Mission Impossible', because of the simultaneous high demographic (arable land per farmer cannot be significantly expanded) and socio-economic pressure (the need for high labour productivity and economic return in the agricultural sector).

Despite heavy investments in both machinery and land-saving inputs, farmers in these countries still face the unpleasant reality of a land and labour productivity that do not match the minimum outputs imposed by their demographic and socio-economic pressure.

The need for high labour productivity (socio-economic pressure) and eventually land productivity (demographic pressure) at the level of society, reflected into changes in technical coefficients at the farm level, is intricately linked to changes occurring in the energy budget of socio-economic systems to which the agricultural system belongs. Economic development, assessed at the socio-economic level, inevitably results in a continuously increasing opportunity cost of human labour and capital at the farm level, and pushes technical development of the farming system towards mechanization, specialization, economies of scale, liberalization of nutrient inputs in the agro-ecosystem, and dependence on price support and other subsidies.

This chain of changes, well-known in the field of agricultural economics though generally described from a different perspective, is the main cause of the abandonment of traditional farming strategies, and the consequent loss of biodiversity in agriculture. Most notably, two important trends follow from this process of technological change : adoption of monocultures, and reduced use of by-products, such as crop residues and manure, in the agricultural process.

Natural ecosystems (e.g. mature natural forest) are able to sustain a large amount of biomass, but with a net primary productivity that is relatively low compared to agro-ecosystems. Traditional farming systems based on gardening in natural forests can produce large quantities of biomass per hectare on the cleared areas, but only with a long cycle of rotation. When the entire area involved in the rotation is considered, overall crop productivity per hectare is quite low.

Other systems of traditional farming based on inter-cropping, multiple cropping and animal husbandry have proved to be highly productive in terms of food supply per hectare whenever ample fresh water is available. However, the combined exploitation of diverse natural resources generally requires labour-intensive manage merit. Rural communities in China and elsewhere in East Asia, masters in the integrated use of natural resources, have been able to feed themselves and pay taxes in form of part of the harvest in spite of a limited endowment of land, for centuries.

However, the particular techniques of multiple cropping, animal husbandry and integrated use of natural resources implied that more than 80% of the population was engaged in agriculture. Moreover, note that the population of China did not surpass 600 million until synthetic fertilizers were made available to China in the late 1970s. Whenever socio-economic development calls for a reduction of the labour force in agriculture, the society must confront the need for an increase in farm labour productivity.

Recycling of natural inputs within the farming system is abandoned as soon as this becomes too costly in terms of land

and/or human labour requirement, provided of course that alternative technical inputs for use in agriculture are available in the market. For example, the trend of abandoning cultivation of green manure in favour of food grain when artificial fertilizer is accessible is discussed in detail by Smil. In general, the decision whether or not to abandon the recycling of natural inputs will depend on the amount of energy or nutrients gained by recycling, divided by the extra hours of labour required, the opportunity-cost of labour, and the price of technical inputs used as alternatives.

When the inputs made available per hour of labour by recycling are significantly less than the inputs that could be bought on the market with the economic return of one hour of labour allocated to off-farm activities, then recycling will generally be abandoned. In this case, linearization of nutrient inputs is more convenient in terms of economic cost/benefit. Within a market economy, the choice to specialize and rely on monocultures often leaves no alternative but to also switch to linearization of nutrient inputs in the form of fertilization.

Criteria adopted for choosing and developing farming techniques have generally been based on socio-economic constraints only (costs/benefits referring to economic growth) and therefore have been biased in favour of continuously increasing densities of harvested biomass.

Ecological constraints imposed by the characteristics of natural ecosystems, that would put an upper limit to the densities of harvested biomass, tend to be ignored. This is for the simple reason that socio-economic constraints to agriculture leading to technological 'progress' have universal application (technical coefficients refer to unit operations in artificial environments and can easily be generalized in terms of monetary costs/benefits), whereas ecological constraints are location-specific and extremely difficult to define and quantify. The consequent imbalance between the short-term and long-term (ecological) perspective when deciding how to manage agricultural production may have disastrous consequences given the immense scale of agricultural activities worldwide approximately 50% of the terrestrial area is devoted to agriculture, while in temperate ecosystems agriculture occupies 70% of the land.

As Hollings has pointed out, in biological systems a general homogenization of activities over a large scale is a disaster waiting to happen: it dramatically increases the risk that perturbation damaging to that pattern will occur and be amplified over the whole system.

In conclusion, strict respect the pattern of natural process in the agro-ecosystem to preserve biodiversity would slow down the throughput of food production per hectare of land and per hour of labour. When these throughputs become so low so as to become incompatible with socio-economic characteristics this solution becomes economically non-viable.

On the other hand, matching demographic and socio-economic pressure above certain ecological thresholds leads to ecological unsustainability. For example, the production of tonnes of grain per hectare per year and hundreds of kilograms of grain per hour of labour with less than 0.2 hectare of arable land per capita can only be obtained in an ecologically unsustainable way through liberalization of nutrient flows in the agro-ecosystem that is continuous harvests (no rotation) and heavily reliance on technical inputs such as machinery, fertilizer, pesticides, and irrigation.

Discussions of sustainability of world agricultural production should carefully consider both ecological constraints such as the possible limiting role of land, water, and other natural resources, and socio-economic constraints to food production.

Current strategies of technological development in agriculture not only appear unable to solve problems of food security in the long-term, but also have been increasingly malfunctioning in the last decade, both in developed and in many developing countries.

In developed countries, this is because excessive socio-economic pressure has caused farms and farmers to disappear, while remaining farming systems employ production techniques that are detrimental to ecological equilibria and in developing countries it is because excessive demographic pressure has caused poverty in rural areas to increase and food production per capita to decrease.

Paradoxically, this evident malfunctioning is perhaps also a reason for optimism. After all, when dealing with human systems, cultural mode-locking is the most persistent and important cause of lack of adaptability in response to changing boundary conditions.

As concerns agriculture, however, there is an increasing, though not yet general consensus among farmers, scientists, politicians and the general public that we somehow have to change the way we look at and practice agriculture. In terms of opportunities for future changes in farming strategies, the situation is dramatically different in developed and developing countries.

This risk-free environment, that was and still is financially supported by the rest of society, has been realized through guaranteed prices, input subsidies, disaster, relief, etc. Indeed, modern high-input agriculture depends economically on the rest of society, that mobilizes economic resources to maintain its agricultural sector. We may thus conclude that in developed countries the economic resources are available to support the agricultural sector, but that they should be spent in a different way.

Developing countries have never disposed over sufficient economic resources to generate a risk-free environment for farmers and investors in high-input agriculture. On the contrary, the agricultural sector is usually taxed to support other sectors of the economy.

In the absence of government support, risks implied by excessive simplification of ecosystems typical of high-input agriculture are transferred to individual farmer families. This makes such a solution less attractive, especially for poor farmers operating in marginal areas. Nevertheless, the continuous increase in demographic and/or socio-economic pressure as well as access to Western technologies, has pushed for the adoption of high-input techniques in local attempts to augment agricultural throughputs. Given the rapidly growing population in developing countries (actual increases in demographic pressure) and the gap still to fill before reaching desired income levels (potential socio-economic pressure), we can expect the density of agricultural throughputs to increase in most, if not all, developing countries, even though this could be destructive for the environment in many cases.

In spite of this wide difference, there is a striking similarity between high-input and low-input agriculture. In both cases, food security totally depends on the availability of fresh water, fertile soil, pollination, natural recycling of nutrients and pollutants, biological pest control and nitrogen fixation. These are services that cannot be substituted by injection of technical capital, and that are heavily dependent on biodiversity.

However, this view implicitly implies the adoption of the entire world as the hierarchical level at which the economic process is assessed and optimized. This level is not necessarily the best possible choice, especially when dealing with agricultural production. In fact, the use of a smaller space-time scale for defining strategies of technological development also has positive effects. First, a smaller scale makes it easier to involve local people in the decision-making process and increase the responsibility of local communities for resource management.

Thus, decisions related to sustainability should be made as close as possible to local people, and with the participation of all major stake-holders. Apart from the ethical aspect, this would make it possible to benefit from indigenous knowledge of the nature of the problem of sustainability dealt with. Second, internalizing most of the external services on which economic systems rely (by reducing trade) makes it easier for the system to respond quickly to changes and signals coming from surrounding ecosystems, inspite of their intricate and varying nature. Societies heavily relying on trade often cannot see the importance of the ecological services on which they directly and indirectly rely. Therefore, energy expenditure in transportation from distant ecosystems should be justified only in cases of substantial comparative advantages. Third, the goal of harmonizing energy and material circulation with the characteristics of local ecosystems (decreasing the environmental impact of economic activities) can be more easily achieved it, the space-time closure of the activities of production and consumption in the society is similar to the spacetime closure of the matter cycles occurring in local ecosystems. These points suggest that the enlargement of scale of economic systems obtained by heavy trading can also induce negative effects, both on the stability of the biosphere and on the integrity of local socio-economic systems. There is therefore also a rationale for policies aiming at increasing the level of self-sufficiency and addressing the issue of social equity at the level of national and regional economic systems.

When defining the terms of reference for sustainable agriculture, biodiversity should be considered as an output rather than as an input in agriculture. Agriculture should not only produce food, but also preserve a healthy environment. The latter task, grossly neglected, involves guaranteeing a higher level of cycling of nutrients, better control of pollution of fresh water supplies, and a sound management of the landscape in general. For all these tasks preservation of biodiversity is essential.

Moreover, the perception of the trade-off between long-term ecosystem health and short-term maximization of production is affected by the level of development. In countries with a high material standard of living, citizens often consider it acceptable to reduce economic returns for the benefit of the environment, while developing countries do not have room to reduce economic returns. They have no other option but to boost short-term returns at the cost of long-term sustainability.

At the moment of discussing agricultural policies in view of sustainability, the small space-time scale typical of economic analysis should always be balanced by a long-term and worldwide view. But is this a feasible idea? Developed countries might be prepared to sacrifice part of their own economic return for the benefit of their immediate surroundings, but doing this for the benefit of other, distant societies and ecosystems are something quite different. The fact that all socio-economic systems, ecosystems and agro-ecosystems form part of a larger whole, the functioning of which is affected by the health of all its individual components, often appears to be beyond our capacity and willingness of understanding. The 'biodiversity revolution in agriculture' should result in a definition of a set of new optimizing parameters for agricultural techniques.

10

BIODIVERSITY LOSS AND THREAT

Data from the last three decades demonstrate that on a worldwide scale, the total area and woodland has remained fairly stable. Considering, the conversion of natural forests in many parts of the world, this means that in general natural forests are being replaced by woodlands which are the result of human activity, be it planting of trees or the development of secondary forests on land which has been cultivated in the past. With the increasing demand for land and for wood worldwide, this trend is likely to continue for decades to come.

Consequently, it can be expected that forests which bear the marks of human influence will play an increasingly important role in species conservation, particularly in tropical countries where the largest biodiversity must be preserved, but also where economic needs put considerable pressure on natural resources. This is being increasingly recognized by forestry officials and forest managers worldwide. Many countries in the temperate zones which no longer have any significant amount of mature natural forest, manage biodiversity in anthropogenic forests.

In many European countries, there is only little need to generate economic returns from those forests, while in the USA, the National Forest Service recognizes a serious need to balance economic demands with biodiversity. In many tropical countries where forests are managed for economic interests, maintaining a diverse forest with a large number of species is often considered contradictory to the principal goal of increasing returns. This, for instance, remains the case with most of the plantation and land development programs that the Indonesian Ministry of Forestry has designed for areas of forest land which have been designated Conversion Forest. Indonesia has a total land area of 199 × 10* ha, including about 10" × 10^6 ha of forest and woodland of which 86.4 × 10^6 ha are forest cover. This is about 6% of the world's total remaining tropical forest, and the third largest area after.

Brazil and Zaire, Indonesia's forests contain about 10% of the world's plant species, 12 % of the world's mammals, 16% of the world's reptiles, and 17% of the world's bird species. Because of its unique geographic location on both sides of the Wallace Line, between the continents of Asia and Australia, a high rate of species endemism occurs. On the other hand, Indonesia's forests constitute an important economic resource for the country. Until recently, timber was the second largest generator of foreign exchange after oil, and remains an important source of development capital, notwithstanding this economic importance. Indonesia has repeatedly expressed an awareness of it's responsibility to the world community to preserve its forests and the biodiversity they contain.

In order to safeguard both the important economic role of the timber industry and the country's biodiversity, Indonesia has functionally divided its large forest. This division was laid down, until 1996, in the Forest Land Use by Consensus which is now being replaced by Provincial Land Use Ordination Plan. Roughly one-third of the country's forest lands are set aside for conservation or protection purposes. The larger part is designated as Production Forest, while the remaining area is designated as Conversion Forest or Non-classified Forests. Conversion and Non-classified Forests may be used for agricultural or other purposes.

Hence, the GOI has made it a goal to eradicate all forms of swidden agriculture in the whole country. So, several programs were designed to reach this goal. These programs vary according to the category of forest in which they are being conducted. So called 'social forestry' programs have been designed to stop forest encroachment by small farmers in Conservation, Protection, or Production Forest areas.

Furthermore, we argue that alternative choices for economic development of swidden agriculturists can be designed which can preserve, or even enhance the biodiversity in these same areas. In particular we will try to demonstrate ; (1) that conservation of biodiversity is an integral part of resource management among the very same swidden agriculturists as an important subsistence strategy : (2) that several of the programs which are designed to eradicate swidden agriculture in Indonesia ironically will lead necessarily to a loss of this biodiversity that is maintained by swidden agriculturists : and (3) that instead of promoting large-scale timber plantation or estate plantations, developing traditional forest management is both an attractive option for regional economic development as well as a way of preserving the biodiversity that is an integral part of many swidden agricultural landscapes.

It is estimated that swidden agriculturists are responsible for about half of the estimated 1×10^6 ha land deforested in Indonesia yearly, although little accurate data exist to sustain such an estimate. Of the total area under swidden agriculture, about 25% is in Production Forests, 15% in Protection Forests, 29% in Conversion Forests, and 31 % in Unclassified Forests. Even though 60 % of swidden agriculture is located in forest land which is planned to be used for non-forest purposes, swidden agriculture is considered to have serious implications for the future of forestry in Indonesia.

Since the early 1970s, the GOI has designed programs to deal with swidden agriculturists. In their earlier phase, these programs attempted to relocate swidden agriculturists out of the forest areas. Especially since the mid-1980s, when the issue of sustainable forest utilization has become widely discussed in Indonesia, many directives from the Ministries of Forestry,

Agriculture, Transmigration, Social Affairs and Home Affairs have appeared which are meant to 'stabilize' swidden agriculture.

The social forestry, or village development programs, which are applied in Conservation, Protection, or Production Forest areas, aim to give people responsibility for managing the forests in their areas. In fact social forestry programs are still ill-defined, and subject to much discussion. Experiences with social forestry in Java, where local control over forests was replaced by government control some time ago, give little hope for similar programs for the outer islands.

Social forestry is not attempted in areas which have been designated as Conversion Forest. In this category of state forest land, three different programs have been defined : (1) the establishment of industrial timber estates; (2) the nuclear estate program and (3) the Rehabilitation and Expansion of Export Crops program. The timber estates development program began in 1984. This program has a high priority with the GOI, since in the near future, it is supposed to substitute for the expected decline in supply of timber as a result of exhaustion of the natural forests. Until 1988, some 1.4×10^6 ha of timber production plantation existed. The largest parts of those timber estates are still found in Java, but the second largest parts are located in the four provinces of Kalimantan with most in West Kalimantan. Indonesia's goal was to reach 6×10^6 ha of industrial timber plantation, by the year 2000. These plantations were to be established in Conversion Forest or Production Forest areas if they do not fall under a forest concession and do not contain sufficient commercial natural forest.

In practice many state HTI projects have been established in secondary forest, scrub land or grass land. The location of HTI estates in secondary forest areas implicitly means that it is located in land that is occupied by swidden agriculturists. One of the objectives of HTI estates is also to rehabilitate degraded lands. However, often any land that is under fallow is qualified as degraded by forestry officials, although from the swidden agriculturists point of view this land still may be productive rice land.

The participation of swidden agriculturists in industrial timber estates is mainly as providers of labour and land. Land needed for the estates is obtained from local farmers who are given some compensation, but this is not higher than US $ 15.00. In general the timber estates were to provide work to the local population, hence to generate income, and thus reduce the need for swidden farmers to grow their own food, and slash more forests.

The wages paid are rupiahs 3750 per day, or US $ 1.70, only when the implementation of industrial timber estates is combined with the Transmigration Program like housing infrastructure, schools, medical posts, and the like, provided to the transmigrants coming from Java, Madura, Ball, and Lombok, as well as local farmers who agree to participate. HTI/TRANS projects aim to include about 50% of 'local' transmigrants in their schemes. The NES/PIR or Nucleus Agricultural Estate Program, which began in 1979, and which is the responsibility of the Ministry of Agriculture, aims at establishing estate crops. In this program government or private companies establish an estate, which may be as large as 50,000 ha. About 60% to 80% of the land is dedicated to small holder production of the estate crop.

Participant farmers are given 2 ha of land to establish the estate crop, and 0.51 ha for a house and a home garden. The estate company provides facilities to the participating farmers to establish their 2 ha of estate crop, and it guarantees to buy the yield, as well as to provide technical assistance and credits for inputs. Farmers accept these facilities as a loan which they have to pay back. In West Kalimantan, farmers must repay US$ 4900 plus an interest of 6%. Farmers who become estate members commit themselves to work as labourers on the company-owned part of the estate, for which they receive daily wages.

In many PIR/TRANS projects local swidden agriculturists are incorporated also, but since the estates rely on transmigrants for their labour, fewer local farmers can be involved which means that relatively more farmers will be resettled. Farmers usually do receive compensation for the land they give up. Reports, however, document the coercive, methods used by the state companies to obtain traditionally held land. NES/PIR or PIR/TRANS estates are entirely converted to mono-crop plantations, except for a smaller part which is converted to food crop area.

In 1993, the total area under oil palm plantation in West Kalimantan, was 1,42,403 ha. The total area planned to be converted into nuclear estate crop is 1.2×10^6 ha, or 8% of the total surface of West Kalimantan. The third program that affects swidden agriculturists in West Kalimantan is the Rehabilitation and Expansion of Export Crops program. This program provides credit to farmers to improve the quality of their small holder plantations. In the outer islands, support is given for 2 ha of export crop land. In West Kalimantan, this program only focuses on improving the quality of rubber production. Very soon after, the first *Hevea* rubber appeared in Sarawak early this century, it also was brought to West Kalimantan by the Dutch.

Most rubber, however, has been grown as 'forest rubber, planted on swidden fields, and left to grow, together with the spontaneously, occurring secondary vegetation. As a result, most rubber gardens are diverse, species rich, anthropogenic forests. Farmers who receive credit and technical assistance under the Expansion of Export Crops Program commit themselves to plant only improved rubber clones. These are high yielding rubber clones, which are much less resistant to pests than the locally used varieties.

In many cases, however, farmers have actually been sold the locally existing rubber to improve the high yielding rubber.

Farmers who plant improved rubber clones are supposed to conduct a thorough slash-weeding from the onset of the plantation until the trees are large enough to be tapped. As a consequence, improved rubber gardens have a characteristic mono-crop species composition which is similar to that of Pinus

Merkusii, *Acacia mangium*, or oil palm plantations and not like the diverse forest rubber gardens found in Kalimantan and Sumatra.

A dominant, mono-culture plantation has a grassy weed vegetation covering the forest floor. Since currently more than 4,00,000 ha of land are under traditional species-rich small holder rubber gardens in West Kalimantan, the extension of loss of biodiversity can expected to be considerable if these areas are to be converted to mono-culture improved rubber gardens.

In 1990, the whole island had 12.7×10^6 inhabitants, and only 9.1×10^6 of which lived in the Kalimantan provinces. About one-third of the people who live on the island are broadly classified as indigenous Dayak. Throughout the history of Borneo the island has been of interest for its natural resources since the pre-Dutch colonial periods for its various forest resources and minerals and for the last several decades for its oil and timber.

Gradually various other groups came to claim their niche : the Chinese, the Buggies from Sulawesi, and later people from other islands from the archipelago, as transmigrants, or as spontaneous migrants, continuously pushing back the native Dayak to more remote places. Together with the booming of timber and oil exploitation came the large scale opening up of large parts of West, South and, too a lesser extent East Kalimantan. This opening up was followed by large scale agricultural development projects. The native Dayak population mostly living in the remoter interiors relies for its daily staple on upland rice cultivation, produced in yearly cleared swiddens. Much controversy and confusion still exists about swidden agriculture.

The term covers agricultural practices which begin with the slashing and burning of vegetation of an area of land, using it for intensive cultivation for one or several years, followed by fallowing the same plot for a longer period. Among most swidden agriculturists, fallow fields continue to have an economic importance since the secondary vegetation contains spontaneously occurring species, or perennial species which, were planted at an earlier stage and which continue to yield useful products.

Some of these species occur mainly in fallow vegetation, and are only exceptionally found in primary forest. Jessup claims that some Kenyan farmers in East Kalimantan purposely maintain a sufficient area of fallow to assure the supply of such species. Hence, as a rule, among swidden agriculturists some kind of management of the fallow vegetation occurs. Among many swidden agriculturists the dominant economic imperative is to optimize returns of labour, rather than returns of land.

Swidden agriculture is still practised by many farmers in the world, and subsequently appears in, many different forms. Even on the island of Borneo, significant variations of swidden agricultural patterns can be observed. For instance, Iban farmers historically had the reputation of preferring land that was newly opened in mature natural forests, and shied away from making swiddens in secondary forest, although Jessup and Sellato question the validity of such generalizations based on ethnic affiliation. Other groups, like the Kenyah cultivated the area around a new settlement for several decades, and then moved the whole village to a new site.

A reason to move the village could be the exhaustion of agricultural land, or because diseases occurred at increasing frequencies, or in some cases the location had become unsafe as a result of invasions by other groups. Recent migrations have occurred because people wanted to gain easier access to markets and government services. The Bidayuh, a name originally used to distinguish the non-Malay, non-Iban inhabitants of Sarawak's first division, constitute several different linguistic subgroups, who originated from the Indonesian part of West Kalimantan, and which are found as far South as the Ketapang basin.

Bidayuh farmers practice a type of swidden agriculture which is more sedentary than that of several other groups in Borneo who migrate more often. Rather than moving entire villages, Bidayuh villages which become too large, are split and only few portions moved to a new area of forest to start their own settlement. The stable Bidayuh villages, however, frequently move the location of their settlement within the village territory, over distances that do not exceed about 1 km.

The more permanent character of Bidayuh settlements has certain consequences for the way swidden agriculture is practised. In order to be able to reside over the long-term in a single region, a farmer must have a sufficiently large holding of land under various stages of fallow. This holding is built up gradually from when a couple become an independent farming household, until the area is large enough such that fields can be used and followed at sufficiently great intervals. Since Bidayuh farmers have more stable holdings, the area to which they hold tenure claims are smaller than, for instance, the ever-moving Than.

The landscape of a typical Bidayuh village is usually characterized not only by land under swidden and fallow, but also by many patches of forests of various sizes. For instance, among two different Bidayuh groups in the North of West Kalimantan, Mate Dayak and Jangkang Dayak, six types of managed forests have been described. These include communally held forest, or hutan tutupan, which are usually located close to villages and may be as large as 100 ha. Hutan tutupan mostly function as a reserve for timber for house construction, rattans, medicinal plants, fishing and hunting sites, and to protect water source. A smaller version of the same type of forest, of between 1 and 3 ha is forest islands or pulau rimba. Such patches of forest become surrounded by one household's agricultural fields or fallows and according to local customary law, became private property of this household. These forests are often kept as private forest reserves, and yield the same products as hutan tutupan. In the same category of protected areas of mature natural forest fall the Sompuat, which are patches of forest which surround honey trees, mostly Koompassia excelsa or Koompassia nialaccensis.

The sompuat forests are kept intact in order to provide an appropriate habitat for the Apis dorsata honey bees that built their nests in the central tree. Obviously, preserving areas of mature natural forests also implies preserving the diversity of species that are present in them. However, that is not where Bidayuh biodiversity forest conservation stops. Two types of managed forests are important, not only economically, but also for the diversity of plant species they contain. These are tembawang or fruit forest gardens, and rubber gardens created by individual farmers planting trees on sites previously used for agriculture or around the settlement area, or sometimes directly in mature forest.

Planting may continue for many years until eventually most of the area designated for tree planting is covered. Such tembawang are privately owned and inherited by descendants of the people who established the forest. They are usually about 1-2 ha in size. Since many farmers in a single village create individual tembawang on adjacent fields, over time these merge into a single stretch of contiguous forest. Because of the short distances over which the village is frequently moved, the tembawang area often trails the advancing settlement as a long band of high closed forest. Rubber gardens are usually located adjacent to tembawang. Once the rubber seedlings are planted, no wedding occurs until the rubber is fully grown about 10 years.

Only then are trails established to reach trees which have to be tapped. Very often additional fruit trees are planted in rubber gardens, as a result of which they tend to become an extension of the tembawang area, rather than a totally different type of managed forest. The biodiversity conservation role of managed natural forests like hutan tutupan, pulau rimba, and sompuat are obvious. However, in this respect tembawang and rubber gardens are no less important.

In one of the villages researched, for instance, 400 plant species were recorded in five tembawang. These five tembawang had an average of 170 plant species per hectare, while one specific tembawang had a total of 231 plant species per hectare. In two plots of mature natural forests in the research area 256 and 258 plant species per hectare were recorded. High species diversity can also be found in rubber gardens.

From the biological conservation standpoint therefore, it would be better to leave swidden agricultural land the way it is, rather than try to implement largescale plantations of mostly single species. However, the objectives of HTI, NES/PIR and NES/TRANS and export crop development are not primarily to conserve biodiversity, but rather to increase or sustain the state's income from timber production, or estate crops.

These incomes are indeed intended to increase living standards of local farmers, but also enhance the country's timber, oil palm and rubber industry, in order to provide income for the stale and to generate foreign exchange. The programs that deal with swidden agriculturists in Production, Protection, or Conservation forests, focus largely on developing alternatives which reduce the reliance of farmers on forests as a resourcc to create agricultural land. Some of the strategies are to set aside forest land from which products can be obtained for sale in local markets to obtain monetary income.

The programmes reserved for areas of Conversion or Unclassified forest land do not consider options of forest exploitation, although they are required to dedicate some land to other uses. For instance, the HTI estates have to dedicate 10% of the land to tanaman kehiduppi, which can be private vegetable or mixed fruit gardens. Similarly, PIR projects provide land to participating farmers for their own house gardens. Considering the objectives of estate development, alternative choices will have to provide the same incomes as the estate options to be socially acceptable. They have to be generated at least partly from some kind of exploitation of forests in order to be able to choose alternative approaches for developing swidden agriculture without losing its intrinsic biodiversity.

Reports exist from Bidayuh farmers in central West Kalimantan who gain substantial income from durian yields from tembawang forests. These fruits are being bought by traders from Kutching, Sarawak who travel for 6 h to reach the producers during the fruiting season. Expanding markets in urban Indonesia, Malaysia and Singapore only will increase the demand for such products.

Hence, the opportunity to increase incomes by producing durians in diverse forests gardens only will increase. Existing forest management practices, furthermore, provide a unique opportunity for small holder timber production which eventually could sustain the supply of the local timber industry. Such timber can be planted in and harvested from rubber gardens, tembawang, pulau rimba and hutan tutupan. The 33% of area under forest in the village where Dayak forest management was studied suggests the potential that such traditional forest management holds for the production of high quality timber.

Some kind of tree cover creates a better environment to plant species which are naturally found in closed forests, rather than in open terrain where many seedlings die, or where fires may destroy plantations. Additionally, in diverse holdings, other faster-growing species could be produced on small lots which would be rotated, for instance, with rice production in order to supply an industry that needs faster-growing species of lower wood quality.

11

SOCIAL BIOTA FOR BIODIVERSITY

Soil macrofauna biodiversity was studied during the May-October periods of the years 1978-1991 in the following set of ecosystems located in the territory of the Central Chernozem Reserve and the Kursk Agriculture Station : 120 ha of natural unmown meadow steppe; 500 ha of mown meadow steppe; 60 ha of pasture, 45 ha of barley field without fertilisation, 1 ha of permanent black fallow ploughed twice every year. The mowing cycle in the mown steppe is hay-making once per year for three consecutive years followed by one year without hay-making.

The meadow steppe and pasture ecosystems are the only grassland ecosystems in the Central Russian region that have never been ploughed during human history. They were used as horse pasture or hay-making land until the 1930s, when they became part of a reserve. Other small islands of real steppe can be found in Ukraine. These ecosystems were thoroughly investigated by climatologists, soil scientists, botanists, geochemists, vertebrate zoologists and other for long periods of time, and a lot of data have been published on the chemical composition and long-term dynamics of the ecosystems.

Plant biodiversity studies have been carried out for more than 60 years while soil animal biodiversity research began about 40 years ago. The main studies on soil chemistry were carried out in 1950-1980, plant chemistry in 1970-1980 in the same ecosystems. In most of our studies we used the following method of soil macrofauna estimation. Soil samples were taken by soil corer up to 30 cm depth.

However, it was found that collection of these large samples led to disturbance of the vegetation cover around every sample of an area four to six times larger than the area of the sample. This sampling procedure could change the vegetation cover in and around places of collection and alter natural processes in the unique reserve of ecosystems. It was therefore decided to modify the sampling procedure, i.e. sample size and number of samples.

Twenty samples of 10 cm × 10 cm every 2 weeks were considered sufficient for the study of seasonal dynamics of soil macrofauna in the reserve habitats. Population number of different species of soil macrofauna obtained by this sampling procedure were practically equal to those found with the earlier method. The total number of soil samples collected during the period of investigation was near 3000. This includes all samples gathered during studies cited below.

The species number and population characters of soil macrofauna decreased in the following order : unmown steppe > mown steppe > pasture > barley field > permanent black fallow. It is noteworthy that in the black fallow ecosystem only one or two species from any of the macrofauna groups were found, namely the earthworm *Nicodrilus caligonosus*, the diplopods *Brachyiulus jawlowskii* and *Megaphyllum rossicum*, the staphylinid beetle *Amischa analis*, and the carabid beetles *Bembidion lampros* and *Microlestes minututus*. None of the geophilids or lithobuds that are common in the other ecosystems were found in the black fallow ecosystem.

Practically, all animals found in the field and the fallow soil are microbivorous or carnivorous. Although grazing pressure at the pasture is relatively low, and along the pasture borders there are patches of steppe vegetation, soil macrofauna biodiversity is low in the pasture compared with the steppe ecosystems. Only the carabids and elaterids show an increased species number in the pasture soil compared with the barley field and black fallow. Amara species carabids and wireworms are new members of the groups under study in the pasture soil, in addition to the species already mentioned for the barley field and black fallow.

Staphylinid beetles found in the pasture soil were all coprofillous. Regarding geophilids, *Pachyinerium ferriigineum* replaces *G. proinnis* and *S. nemorensis* found in the barley field soil. Density and biomass of the soil macrofauna population in the pasture soil is similar to that in the soil of mown steppe, but biodiversity is lower.

Biodiversity is notably greater in the steppe ecosystems compared with the other ecosystems studied and is comparable in the mown and unmown steppe plots. However, density and biomass are markedly higher in the unknown steppe than in the mown steppe. There are no differences in the species composition of earthworms, diplopods, geophilids, lithobiids and elaterid larvae identified in soils of the mown and unmown steppe. Some new species appear in these groups in addition to those mentioned above.

The differences are marked only for population density and biomass of separate species in these groups. *E. nordenskioldi* is the dominant species in the unmown steppe soil and *N. roseits* in the mown steppe soil. On the other hand, the staphylinids and carabids differ markedly in these ecosystems. The Soerensen index for carabids is 0.64 and for staphylinids is 0.52. The Jaccard index is 0.47 and 0.36, respectively. All other community parameters, such as mean species number per sample, mean density and biomass of animal populations, are correlated with the changes of biodiversity. Mean species number per sample shows a stronger correlation with the population density and biomass than those total species number. This is because mean species number per sample is a characteristic of dominant and common species in determined density and biomass.

A decrease of biodiversity of soil mesofauna groups is a common phenomenon in sets of ecosystems with increasing human influence. Among the many possible reasons, soil perturbation, and microclimatic and soil chemistry changes are considered the leading causes. Even a low-impact human activity such as hay-making is reflected in the soil macrofauna composition and differences between the mown and unmown steppe communities are considerable.

Pitfall data collected in the same period indicates that in the mown steppe, the share of ubiquists among staphylinid beetles and carabid beetles is 17% and 5% respectively, in the unmown steppe it is 28% and 7%, respectively, in the pasture 32% and 44%, respectively, and in the barley field and black fallow ecosystems 100% for both groups. It is interesting that the share of ubiquists in the field increased from 72% in the 1960s to 100%. This may be a resit of an increase in the human impact on the local environment as a whole. In our opinion, the higher proportion of ubiquist species of beetles in the unmown steppe is related to the microclimatic conditions in the thick layer of litter, where conditions are optimal for ubiquist species.

On the other hand, *E. nordenskioldi* is the only non-ubiquist species among earthworms in both ecosystems but is dominant in the unmown steppe. Two other species are found in all anthropogenic habitats and may be considered as ubiquists with *N. wscus* dominating in the mown steppe. Based on data on biodiversity, it is difficult to estimate what type of a reserve regime is closer to the initial prehistoric conditions of real steppe (mown or unmown steppe). The grass litter layer is very thick in the mown steppe but practically absent in the mown steppe.

Geobotanists consider pasture regimes to be the initial one for real steppe, and the mown steppe regime as analogous to the pasture (grazing) regime. However, it is evident then that pasture and mown steppe cannot be considered as analogues. The biodiversity and community characters prove this assumption. Grazing activity in pasture at the level of one cow per hectare is sufficiently high to disturb steppe ecosystems and change them entirely. It is very significant that this level of pasture activity does lead to the disappearance of the grass litter layer.

Many pasture soil species are ubiquists or coprofillous. Our observations showed that such coprophagous beetles as Aphodius and Onthophagus species are common in surface soil of pasture. These species are practically absent in the mown and unmown steppe. The point of view that pastures and mown steppe are analogous ecosystems was universally adopted by geobotanists. However, a study of species composition at a steppe plot after artificial fire shows that during the first 2 years soil fauna species composition is similar to that on mown steppe plots.

Then species composition is changing toward that typical of unmown steppe. It confirms our opinion that steppe ecosystems are cyclicphyrogenic ecosystems, and not pasture ecosystems. Therefore the species composition of soil macrofauna in the mown and unmown steppe denotes that these ecosystems can be regarded as separate stadia of post-fire succession series in meadow steppe, and not as an initial state of real steppe ecosystems. Earlier we showed that changes in biomass and production of soil animal populations in agroecosystems depend on available organic matter and nutrients in arable soils. Correlation between species composition and population density and biomass suggests that resources of nutrients also have an influence on biodiversity.

The differences in composition and biomass of the soil fauna in arable plots are small although the quantity of organic resources in the field soil are somewhat higher than in the fallow soil. In the fallow soil there are no such sources of available carbon, nitrogen and phosphorous as plant roots for animals and microorganisms. Available organic matter for animals in field and fallow soils is mainly present in the form of decomposable humid compounds and microorganisms.

The low biodiversity in field and fallow soils is probably a reflection of small pools of available organic matter, nitrogen and phosphorus sources for the fauna. Stores of available organic matter in the natural steppe are three times higher than in the field and five times higher than in the fallow. The microgen pool in the steppe is 1.5 times higher than in the field and fallow. The phosphorus pools in the different ecosystems are practically equal.

However, microbial biomass in the soil and in the tits of soil animals are the main sources of essential nutrients for animals. Biomass of microorganisms in field and fallow soils is higher than in meadow steppe soil, but it is only 'potential' biomass. To be the nutrient pool for animals, microorganisms need to store the available energy in soils. These stores are concentrated in dead plant residues and plant exudates.

Hence, the biodiversity of soil fauna is possibly correlated with amounts of available organic matter and nutrients, and not only with habitat conditions. In arable soil it will be dependent on input of available carbon compounds for soil microorganisms. If we compare overall biodiversity of ecosystems we usually only consider biodiversity of multicellular organisms and not the biodiversity of microorganisms. Unfortunately, there are some problems with the comparison of biodiversity of taxa of different size levels, and of ecosystems as a whole. Over 10 years ago, we first suggested a new index for biodiversity comparisons.

Our aim was to find indicators for a more objective estimation of biodiversity. In our opinion, DNA in ecosystems could be an indicator of 'potential' biodiversity in an ecosystem as a whole. We called the suggested character the 'index of genetic saturation' (IGS). IGS is equal to the sum S(mjx kj), where S is the sum of the products of mj and kj for species of a separate group (S is equal to the species number in the group) in an ecosystem.

12

CONSERVATION OF BIODIVERSITY IN FUTURE STRATEGIES FOR INDIA

If you plan for one year, plant rice.
If you plan for ten years, plant trees.
But, if you plan for one hundred years.
educate the people.

Development, both individual and societal, and environmental are two sides of the same coin. Due recognition is needed for global environmental risks while designing any environmental strategy, environmentally and socially disruptive economic behaviour must be changed if we intend to p. Livide discarding environmentally sensitive forms of development can only bring about decent livelihood to all citizens of the present and future generations and this change. Rather, the stress should be on qualitative development based on efficient use of energy and natural resources, recycling of end products and minimizing the use of non-biodegradable products.

In a country with a burgeoning human and livestock population, concrete actions and sound strategies are easy to formulate but difficult to implement. Human beings have a tendency to prefer present goods to future goods and in the process, overlook the side effects. To sustain 1.5 billion people in India by the year 2025 heavy financial investment and state-of-the-art technology will be needed in agriculture, water resources, infrastructural development and so forth. Ecological balance will definitely be adversely affected to create liquidity despite our best efforts not to do so. Primary and secondary forest Industries will use more and more wood, cork, bamboo, cane, gum, resins, essential oils, seeds, medicines, etc., for home consumption as well as export in a liberalized economy.

Despite all adversities one has to adopt viable strategies for the conservation of species as this will only guarantee our future and lead us towards the goal of sustainable society, sustainable economy, sustainable use and sustainable development. The strategy should be aimed at policy makers, all involvement and Non-Governmental organisations. This will help in achieving the objectives of biodiversity conservation and sustainable development. Any conservation measure has to be carefully planned so that:

1. Environmentally and socially disruptive behaviour is altered to provide a decent living for all.
2. Production pattern is socially equitable.
3. A human centered, knowledge intensive civilization is built.
4. A balance is maintained between all forms of capital i.e. human, natural, physical and financial.

Some of the measures suggested as a part of long-term planning are as follows :

Carrying Capacity of the System

Carrying capacity varies from region to region depending upon population pressure, their dependence on food, water, energy, raw material requirements, waste production and import and export. In order to live within the carrying capacity the society will have to change the life style and consumption pattern in such a manner that causes least damage to the ecosystem. Over population in a particular region has to be avoided and excessive pressure on land must be reduced. It calls for changes in the prevent land use policy, overhauling of land tenurial system and population control. Forest land will have to be used in such a way that maintains their biodiversity, productivity, regeneration capacity, vitality and their potential to fulfil, now and in future, relevant ecological, economic and social functions.

Enhancement of the Protected Areas

One of the best means of protecting the natural forest is through the establishment of protected areas, an effort that has taken more than a century to evolve from the concept of reserve forest to national parks and biosphere reserves. These protected areas serve three basic functions viz., benefit, education and recreation of the people.

Although, India has made steady progress in the management and expansion of the protected areas, yet, the problems are being compounded with time. More establishment of protected areas does not ensure socio-economic development and species conservation. The management has to take timely action for the speedy disposal for the settlement of problems. Local people who consider themselves as the primary stakeholders must be involved in the use and management of the resources. Socio-economic well being of the people should be one of the top priorities for minimizing the resources dependency on the core of these protected areas. Biological resources in the protected areas can be assigned for consumptive and non-consumptive values so that the conservation of biological diversity can be linked to sustainable socio-economic development.

At the same time, there is a need for change in the attitude of the field staff towards the interests of the local people. This conflictual relationship has to be transformed into cooperation and partnership. Field staff will have to appreciate that with inadequate work force prohibitive management strategy is impossible to implement. Therefore, in the long run a conciliatory, interactive and motivational approach seems to be the best.

Enhancement of protected areas will need effective management with clear objectives for each area. This will require increased investment in human resources, education and research.

Stress on Integrated Development

Of late, it has been realized that a direct stress on conservation measures and neglecting the economic development of the people in and around the protected areas leads to impoverishment and initiates the vicious cycle of subsistence, emigration, land fragmentation, degradation and deforestation. According to the new concept, people will have to change their consumption pattern, and find alternatives on their own, once their purchasing power increases. The only fear in this integrated effort is that short-term conflicts may arise in the absence of appropriate policy measures causing hardships to the people. It is therefore essential that conservation, rural development and compensatory incentive programmes should be organised simultaneously.

Research Impetus

There are five main areas, which need urgent attention:

Forest management

Forest management has undergone rapid changes in the decades of eighties and the nineties. Geographical Information System and Remote Sensing Techniques have the potential to provide a large amount of information on forest resources, growth and biodiversity.

Restoration of degraded lands

Research efforts should concentrate on enhancing the soil microfauna and fertility status so that sustainable production can be achieved by adopting agroforestry systems that can support the economic development of the region. Efforts should be made for genetic improvement and dissemination of multipurpose tree species that can grow in degraded areas.

Research in harvesting, processing and marketing

There is a need to develop low impact harvesting systems, drying and conversion processes for non-commercial abundant species, especially the NTFP's Marketing management has been one of the weakest sectors of forestry and needs new directions and support.

Policy, institutional and socio-economic research

Socio-economic objectives are best pursued if the policies regarding land use rehabilitation of degraded lands, benefit sharing, land and tree tenure, participatory planning and resource management are appropriate to the sustainable management and biodiversity conservation.

Research in biotechnology

Sustainable development and biodiversity conservation in the 21st century will depend upon proper exploitation of economic potential of genetic diversity using biotechnology for the production of goods and services required by the increasing population. Foresters will have a dual role in future; one as in situ conservationists and the other in ex situ maintenance of biodiversity for commercial use.

Last but not least, the concept of conservation and development should permeate our thinking and perception. We are all stakeholders, participants and decision makers in this struggle and can achieve the objectives by our concrete actions, strength and determination with the active support and participation of the people.

Strategies, Opportunities and Options

In India, creation of Sanctuaries, National Parks and Biosphere Reserves has been adopted as the major task of achieving the objectives of conservation. Restricting human interference to conserve biological diversity is an act opposed by the rural tribals already facing complex problems, including small land holdings, meager employment opportunities and, over and above, natural hazards, reduced productivity of jhum fields and environmental uncertainties. Conservation strategies must address to the needs of posterity but, at the same time, should not neglect the present needs of the people.

Biodiversity conservation strategies and goals, should redefine biodiversity in the context of local needs, usages and values as priority of conservation. The current experience in the region is that there is lack of objectives of biodiversity conservation strategy, which need to be explicit, clear and achievable. It should not neglect the anthropocentric objective addressing the local peoples' priorities to provide, if not enhance, sustainable and economic benefits.

Current policies and practices of exclusive protectionism of conservation areas need a re-evaluation. After all, people and their culture have historically played significant role in conservation and management of what was there in the past and what is remaining now. There is the need to survey peoples' needs and attitudes around our conservation areas. Possibilities of controlled harvesting or sustainable use of at least certain wildlife resources and minor forest produces (MFPs) by the communities neighbouring protected areas should be considered. Activities that may be taken up as pilot projects may be identified. Such objectives may also ensure long-term survival of wildlife and biodiversity conservation, since the people themselves would want to protect the protected areas. On the other hand, damage to property and human life by wildlife, and the amount and system of compensation payments are some of the issues that are not yet well documented.

Many of the protected areas in India are indeed forests for wildlife shelter in the midst of intense human pressure and activities. Such islands of shelter cannot always afford to accommodate many of the wild animals who possess migratory instinct. There is, thus, the urgent need of identifying animal corridors and bringing such corridors within buffer zones or some kind of management practices in order to reduce man-animal conflicts or wildlife depredation.

In India, it is a fact that there exists rich biodiversity even outside the PAs, such as in sacred groves, traditional village/community/individual forests, in cultural jhum fields and fallow lands. Such Lands/forests are under the ownership/control of the Village Councils, District Councils, traditional chiefs or private individuals. These traditional socio-political institutions must be taken into confidence while formulating strategies for biodiversity conservation. Management strategies and policies should ideally encompass the areas under the control of these institutions, which can be achieved through appropriate dialogue and negotiations without infringing on their land rights/ownership.

There is ample scope for amalgamation of traditional practices and modern scientific knowledge for solving the problems of PAs and biodiveristy conservation in the region. Not that all traditional societies are always wise stewards of biodiversity or their practices are scientific, conservation oriented or sustainable. Such practices should also be identified and discouraged and alternatives evolved.

One of the weaknesses of current conservation strategy in the region is its heavy emphasis on conservation of wilderness areas of forest lands. Conservation of cultivated crops, ethnobiological species including medicinal and wild edible plants is a neglected frontier. This weakness must be rectified immediately as the region has a variety of food and medicinal plants, many of which are not much known outside the region and may have potential of providing alternatives for food and medicines in the region and elsewhere. Such crops have also been reported to be pest resistant and stand long storage.

In various silviculture practices and compensatory plantations in the region, there is often a tendency to patronize one or more better performing species such as pine in the higher altitude and teak in the lower elevations. Such practices should be reviewed and minimized, as mono-cultures reduce the spectrum of biodiversity, both of plants and animals. Ideally, there should be an assessment of areas that may be covered under mono-cultures and native mixed species should be planted to maintain biological diversity.

The development projects in India are not sensitive to biodiversity issues. Many of them are harmful to biodiversity to a greater or lesser extent. Therefore, there is a necessity of improvement in project appraisal methodologies and implementation of programmes with a view to making them more sensitive to biodiversity, while at the same time, ensuring sustainable development, better valuation of renewable resources, improved forests/wetlands/wildlands conservation and increased agricultural productivity. There is also the need for sensitizing the project authorities towards biodiversities awareness and initiatives.

In India, the Environmental Impact Assessments (EIA) in any region tends to concentrate on local physical, biological and ecological impacts, their effect on the human population and local environment. However, their effect on the biodiversity is invariably neglected. One of the main practical constraints to improve EIA is lack of data and the difficulty of ascribing values to all but direct use-values of biodiversity.

Inter department approach

There is also a lack of inter departmental approach and vision in relation to biodiversity conservation in the region. For example, while the state Environment & Forests/Wildlife Department may be enthusiastic to conserve wetlands, the Fishery Department may like to promote commercial fish farming in such wetlands, or the Agriculture Department may like to reclaim the swamp areas for wet rich cultivation. Moreover, there is often random introduction of exotic species or hybrid varieties for short-term commercial gain endangering patronage of local species or varieties. Such practices also dilute the indigenous knowledge systems or erode germplasm of the native species.

Urgent Need

There is an urgent need for an increase in research activities, information processing and dissemination system, either to improve priority setting, or to guide management of biodiversity conservation, both within and outside the protected areas and even for almost all habitat types. Lands and resources may be allocated to the best possible effect. There is also the need for chronicling traditional natural resource management systems, which should be evaluated, strengthened and encouraged. Such knowledge should be exchanged or transferred between tribes for higher ecological and economic efficiencies, particularly conservation and management of biodiversity under their control.

Environmental Education

Environmental education, awareness building and motivation initiatives on biodiversity issues using local arts, cultures and languages should also be a part of the strategy. Ideal traditional plant and wildlife knowledge and skills should be identified, documented and used as educational tools. There is also the need to infuse traditional values, cultures, skills and knowledge into the young generation, many of whom are now exposed to the mainstream life and culture, for better appreciation of their own native wisdom and resources. Environmental education, particularly awareness in relation to implication of biodiversity trade, conventions, regulations, removal and extraction of endangered species, etc. should also be extended to the law enforcing agencies.

Future Strategy for the Conservation of Biological Diversity

The rate of growth of human population must be reduced. Greater emphasis should be given to better management of available biological resources rather than to just consider conservation of biological diversity in isolation. The reason is that increased production of fuel wood, fodder, building material, etc., in areas poor in biodiversity in the country is essentially required to offset the pressure on biodiversity rich areas. A natural resource policy may be formulated to achieve four basic goals:

1. Maintenances of adequate resources.
2. Conservation of resources through reduction in demand and achievement of greater end use.
3. Maximum use of renewable resources.
4. Reduction in dependency of non-renewable resources.

Pic. 28.1. Conservation of Biodiversity

13

BIODIVERSITY AND NEOTROPICAL PRIMATES

Traditional classifications divided New World primates into "marmosets" or "callitrichids" and "nonmarmosets" or "cebids", in which the marmoset patterns of morphology were seen as primitive and those of cebids as advanced. Most authorities recognized 16 extant platyrrhine genera, including Ateles, Brachyteles, Lagothrix, Alouatta, Pithecia, Chiropotes, Carajao, Aotus, Callicebus, Cebus, Saimiri, Cebuella, Callithrix, Leontopithecus, Saguinus and Callimico.

In contrast with the general concordance of reviewers working at the genus level, considerable disagreement exists among authorities working at higher taxonomy levels. In a 1986 revision, Ford reorganized recent platyrrhine genera using a cladistic approach. In 1992, Rosenberger of tin-National Zoological Park, Smithsonian Institution, launched a new synthesis based on phylogenetic relationships, adaptive change, and the fossil record. He recognized four monophyletic subfamilies in that classification, Callitrichids became callitrichines, closely related to squirrel monkeys and capuchins. Many features of callttrichines are now thought to have resulted from selection for small size and thus are derived and not ancestral or primitive traits. Identification of the phylogenetic link between the Cebinae and Callitrichinae was made possible by the discovery of a new set of fossils at La Venta, Colombia, with morphological characteristics intermediate between those of the two subfamilies.

The rate of description of fossil platyrrhine genera has increased markedly in recent years. Of 16 known fossil genera, 7 were described since 1985 and 4 of these since 1990. Our understanding of phylogenetic relationships and extinction of neotropical primates certainly will continue to improve if the trend toward increased description of fossil forms continues. New advances in molecular genetic techniques now make it possible to objectively test hypotheses about the phylogenetic relationships among the platyrrhines.

The Brazilian geneticists Maria Sampaio and Horacio Schneider examined sequences of DNA in nuclear globin genes and found close relationships between Alouatta and the atelines, between Cebus and Saguinus, and among Pithecia, Chiropotes, and Cacajao, adding additional support to the phylogenetic model developed by Rosenberger. Much of what we know about the diversity of neotropical primates, at the species level is due to the painstaking descriptive detail provided in the published works of Hershkovitz.

The biological accuracy of taxonomic designation becomes particularly important because legislative protection and conservation funding for full species is significantly greater than those for subspecies. In addition, conservation objectives of reintroduction and translocation of isolated populations are predicated on the conservation of genetic diversity within a species. If genetic management is to have imperative that we be able to distinguish conspecifics forms from less closely related forms.

Revision of neotropical primates at the species level should be given high priority, both in terms of contributions to primate conservation and basic search. Recently developed noninvasive genetic techniques such as sequencing of mitochondrial DNA extracted from hair follicles of museum may make it possible to resolve some of these controversies.

Living neotropical primates are found from central Mexico to northern Argentina and include some 68 species, comprising about one-third of all living species of primates. In the mid 1980s, much emphasis was put on identification of major phytogeographic regions that contained relatively large numbers of primates, particularly rare, endangered, or endemic species. These analyses resulted in maps of priority areas for primate conservation and research. The discovery of four unknown primates in such a short period of time, including one species in the backward of one of the largest cities in the world, underscores how little we really know about the diversity of even well-studied taxa such as nonhuman primates and emphasizes the need for more basic research and surveys on primates in the Neotropics. However, if conservation efforts on behalf of any species of primates are to succeed, it will be necessary to go beyond the identification of its geographic distribution and taxonomic relationships with other primates. Although biologists have been conducting censuses of neotropical primates for decades, lack of standardization of methods and the difficulty of observing primates in closed forest has made it difficult to assess the accuracy of the results of these surveys.

Nonetheless, numbers of neotropical primates appear to be decreasing. For example, in 1969, the Brazilian conservationist Alvaro Aguirre estimated that 2,000-3,000 muriquis survived in Brazil's Atlantic forest. In the mid 1980s, the minimum number of muriquis known to exist in 10 locations as published as 240. However, it was not until 3 years ago when Kicrelff, a Brazilian graduate student, used playbacks of golden lion tamarin vocalizations in virtually every forest that might contain these primates—that an accurate estimate of the number of individuals in the wild became known.

Thus, the first exhaustive census for any neotropical primate, suggested that only 559 individuals in 103 groups remain in the wild and that the species is subdivided into 16 isolated subpopulations, 12 of these containing only 1 group. The accurate nature of this information removed any question about the degree of threat to this species and prompted significant changes in the focus of conservation and related research activities.

Monitoring changes in quantity and quality of habitat

In 1986, critical regions for conservation of neotropical primates were identified as huge phytogeographic areas such as Amazon basin and "megadiversity countries" that contained a large diversity of primate taxa. Since that date, our ability to pinpoint regions for conservation has improved markedly. In 1990, approximately 100 specialists on Amazonian fauna, flora, soils, and climate met in Manaus, Brazil, to produce a map of biologically important regions in the Amazon Basin based on criteria of endemism rarity and diversity.

Important here is the diversity of specialists contributing information and the speed with which the results were disseminated. Perhaps the most significant recent advance in our ability to quantitatively evaluate primate habitat is the combination of remote sensing technology and geographic information system (GIS) analysis. A decade ago satellite images of regions of South America were expensive, difficult to acquire, and could be analysed in only a very few locations.

Currently, a variety of high-quality inexpensive magnetic image data is available from several sources and cover virtually every region on Earth. Recent improvements and reduction of handheld geographic position system (GPS) equipment now allow a researcher to precisely determine latitude and longitude for samples of habitat, this simplifying the ground-truthing of satellite data. Desktop computer versions of GIS software make it possible to conduct sophisticated habitat analyses without acquiring specialized equipment.

New techniques such as gap analysis, which evaluate the protection status of plant and animal communities by GIS overlay of distributional data on maps of existing protected areas, help conservationists to be proactive rather than reactive in their efforts to preserve biodiversity. The accessibility of satellite data and new analytical techniques resulted in three large steps forward in terms of establishing conservation and research priorities in the Neotropics.

First, we now are able to quantify changes in vegetation patterns over time. For example, in the 1980s, widespread disagreement over rates of deforestation in the Brazilian Amazon hampered efforts at international collaboration in conservation. In the 1990s, independent reports by Brazilian and American scientists agreed that previous estimates were several-fold larger, than actual rates of deforestation.

Since future treaties on biodiversity and climatic change are likely to require countries to limit their emissions of greenhouse gases, including those resulting from deforestation, estimates of deforestation based on satellite data acquire additional political and economic importance. Second, we now can use satellite data to test hypotheses about the effects of anthropogenic factors on habitat quantity and quality. For example, we can use archival data to document habitat changes that followed the construction of a road into a frosted area in the Amazon and subsequently use that information to predict how a proposed road might alter undisturbed forest in another region.

Third, based on any criteria that can be mapped, it is now possible to quantitatively evaluate areas for creation as reserves, or as corridors between reserves. Remote sensing and GIS techniques are becoming increasingly powerful research tools and should be included in the tool kit of any researcher or conservationist doing field work on neotropical primates.

Ecology and Evolution

Hundreds of scientific reports have been published in the past decade on the ecology and behaviour of neotropical primates. Field research using the comparative method, long-term demographic studies, and research addressing the effects on human disturbance factors on primate population, size and structure. Although the principles of evolutionary biology should be useful in explaining primate behaviour, primates rarely have been at the forefront in this work. Cheney *et al.* identify several reasons why primates are rarely used to test evolutionary theory.

Most species of primates are difficult to identify individually or even to observe under field conditions. Many species are so long-lived that accumulation of data on life histories may require decades of observation. In addition, the social behaviour of many species is sufficiently complex that it becomes difficult to collect information on a sufficient number of individuals to obtain a complete picture of patterns of social organization.

Seasonal and annual variation further complicates the design of field studied examining the effects of ecological variables or the adaptive significance of specific behaviours in tropical ecosystems. Finally, formulation of generalizations often is impeded by differences in researchers, methodology. One approach that has proved useful in circumventing these obstacles is to ask how several primate taxa behave under the same circumstances and when the same observational methodology is used.

Conclusion then may be inferred from differences between species. This comparative approach has the added advantage of focusing on an interesting question rather than on an interesting primate. A good example of the comparative method applied in field research on primates is Terborgh's long-term study of the comparative ecology of five New World primates in Man National Park, Peru.

Comparisons across species allowed Terborgh to make inferences about, for example, the function of territorial defence, factors that influence size of the group and the relationship between size of the group and the social system in primates. Another fruitful comparative study was conducted by Peres at a remote site in Amazonas, Brazil. Peres compared the feeding ecology and habitat use of 13 primate species for a period of 20 months.

Parallel studies of native and reintroduced tamarins allowed researchers to generalize about the effects of captivity on locomotion, communication range use foraging and reproduction. A second approach that has proved useful both in terms of basic research and conservation application is long-term monitoring of demographic variables for a single primate population.

The conservation implications of these data are discussed below. Calculation of reproductive success under field conditions can be problematic, particularly for species of primates in which more than one male copulates with a reproductive female. Under these circumstances, new molecular genetic techniques, notably DNA fingerprinting, may allow paternity excision and thus direct determination of male reproductive success. It has been long thought that deforestation can have negative effects on most primate populations. However, the effects of these and other anthropogenic factors rarely have been quantified for neotropical species.

In recent reviews of the impact of habitat disturbance on a variety of species of primates, Johns and Skorupa reported "that body size and degree of frugivory were negatively correlated in species" probability of survival. Redford and Robinson surveyed the available literature and predicted that hunting by humans would cause primate population density to decrease as prey body mass increases. Preesc *et al.* surveyed primate densities in Bolivia and Peru and suggested that predation by humans was the most important factor affecting primate densities.

Peres used transact censuses of primates in hunted and unhunted areas to test hypotheses about the effect of hunting on primate densities. Peres found that body size alone largely determined the choice of primates hunted in Amazonian forests and that large species, such as woolly monkeys, often were hunted to extinction in areas near human settlements, even in the absence of habitat disturbance. Redford reached similar conclusions for a variety of large animals in neotropical forest. Two conclusions are suggested by these studies.

- Conservation efforts need to focus on large-bodied primates with low rates of reproduction, even before habitat destruction begins.

- Past and present hunting pressure must be included in population dynamics of large-bodied primates, even in areas of apparently intact habitat.

Conservation strategies

The alarm that sounded in the mid-1980s about the immediate threat to most species of neotropical primates was answered in a number of ways. I have classified these strategies as information management and networking application of scientific method, captive breeding, conservation education and international collaboration. The advent of widespread use of electronic mail dramatically increased the speed and decreased the cost of international communication.

Transfer of data files, even from remote field stations, has become routine in the past years. Computerized bulletin boards are available on a number of relevant topics, including conservation biology, primates, ecology, the Neotropics in general, as well as several tropical countries. Many international nongovernment conservation organisations (NCOs) support regional programs to develop and conduct conservation and activities in neotropical countries.

Some of these initiatives, e.g., the Golden Lion Tamarin Conservation Program in Brazil, which has been supported for 10 years by grants from the World Wildlife hind and other NGOs, focus on conservation of primates and their habitat. The Nature Conservancy sponsors the establishment of Conservation Data Centers in 3 Latin American and Caribbean countries to aid in decision-making in conservation and sustainable development. Conservation International supports a Rapid Assessment Program in which teams of experts census diverse taxa of fauna and flora to set priorities for conservation action.

Scientific method

The past decade has seen a heartening increase in the generation of conservation theory and the application of rigorous scientific method and problem-solving techniques to conservation questions. Graduate degrees in conservation biology now are offered at many universities in the United States and Europe.

The National Science Foundation now funds grant proposals in conservation and restoration biology. Although the primary concern of conservationists is the survival of ecosystems, much of the theoretical and applied research in conservation biology is conducted at the population level. The link between population and ecosystem conservation is thought to be the appropriate selection of population of "flag ship species" that serve as the focus of conservation campaigns, "umbrella species" that have ranges large enough to include significant portions of natural ecosystems, and "keystone species" whose presence or absence determines the abundance and distribution of a number of other species.

Computer models such as VORTEX are used to estimate the probability of population extinction and time to extinction based on the effects of stochastic events such as inbreeding environmental variation and catastrophes on population demographic variables. The probability of survival of multiple subpopulations in a metapopulation also may be estimated.

Examples of PHVA applied to four species of lion tamarins are presented in Seal *et al.* The advantages of PHVA include the simultaneous treatment of several variables, the collaborative process by which the best available information is assembled for all interested parties, the quantitative nature of the results, and the versatility in evaluating management options. The shortcomings of PHVA are related to the assumptions implicit in the computer model and the effects of estimating unknown values.

For example, the impact of environmental variation on fecundity and survivorship rarely is known for population in nature. Small changes in the specified effects of environmental variation on reproduction or mortality can produce 10-fold differences the predicted probability of extinction of the population.

Captive breeding and reintroduction

Not long ago, zoos were seen as consumers of primates and other animals, competing to assemble a collection of the rarest and most valuable examples for public exhibit. Recent changes in the objectives of modern zoological parks make these institutions important contributors to in situ conservation of neotropical primates. The principal objective of captive propagation now is seen as the re-establishment or reinforcement, not replacement of wild populations. With over 13,000 specimens in zoological collections, primates are overrepresented relative to other mammalian orders.

Nevertheless, only about 50% of primate taxa are represented in captivity. International management committees, organized to maximize retention of genetic diversity and minimize probability of lion in captivity and the wild, are increasing for many vertebrate taxa, including primates. A number of advisory groups work to coordinate the ex situ and in situ primate conservation activities of zoos.

In 1992, a Tax on Advisory Group (TAG) to the American Association of Zoos and Aquaria (AAZPA) was formed to coordinate the facilitate captive breeding of New World Primates in support of in situ conservation efforts. Paunal Interest Groups (EIGs) coordinate conservation activities in zoos for fauna and flora from a specific region, e.g., Brazil, Studbooks, containing complete demographic records for a species ill captivity, facilitate the scientific management of captive populations, e.g., by facilitating selection of appropriate individuals for pairings reduce loss of genetic diversity. Studbooks are maintained for 15 species of neotropical primates, and have been proposed for several others. For a few species, information on native populations is sufficient for inclusion in studbooks.

The reintroduction of captive-born neotropical primates has been initiated for several species in the past few years. In addition to the reintroduction of golden lion tamarins mentioned above, reintroduction and monitoring of *Ciillithrix geoffroyi* in Espirito Santo state, Brazil, began in 1991. The costs and crik-rui for success in a primate reintroduction are reviewed by Kleiman et. al. The IUCN-SSC Reintroduction Specialist Group and the AAZPA Reintroduction Advisory Group both were created in the past few years to facilitate in situ conservation through integration of wild and captive population management.

Conservation Education

The role of community education in directing the altitudes and behaviours of local peoples toward sustainable use of natural resources long has been recognized as an essential component of any viable conservation strategy. This is particularly true where charismatic species such as primates can be used as flagships for conservation efforts.

Whereas 42.5% of these responses were attributed to project activities, only 6.9% reported seeing tamarins in the forest were mentioned most often included television and radio, a total effect roughly equivalent to all other project activities combined. Graduate training in conservation education for South and Central American nationals is one of the best investments that can be made to ensure the future of neotropical primates.

International Collaboration

A final example illustrates progress in international collaborations. In this case, private lands were purchased and annexed to a Brazilian federal reserve too small to guarantee the survival of an endangered primate. A 1'HVA for golden headed lion tamarins, conducted in the Una Biological Reserve, southern Bahia State, Brazil, suggested that the probability of extinction of that primate was relatively high. Based on the findings of the I'HVA, a coalition of eight international conservation organizations conducted fund-raising efforts to buy out squatters and purchase land to annex to Una Reserve.

Funds were transferred in several instalments to Fundacao Biodiversities, a Brazilian NGO, which purchased lands at market price from local landowners and immediately conferred title to the Brazilian agency responsible for administration of the Reserve. To date, the area of the Reserve has been increased from 5.342 hectares to 7.059 hectares.

14

BIODIVERSITY FOR TROPICAL

Life on earth takes many forms and comes in all sizes, from microscopic one celled plants to blue whales and human beings. Together these organisms and their interactions constitute our planet's biodiversity. Among this profusion of life are the beetles and their insect and arachnomorph relatives, which, taken together, constitute most of Earth's biodiversity. There are 1.4 million species of insects described in the scientific literature, which is about 80 per cent of all life currently recorded on Earth. Taxonomists, those who named and classified species, had been describing species of insects at about 4,400 per year for more than 235 years, and in the last 25 years, they had described about 8,680 species of insects per year (±363). This written record is at best perhaps only 3.4 per cent of the species actually living on the planet. Recent estimates of insect species, mostly in tropical forests, indicate that the descriptive process is woefully behind. These estimates indicate there may be as many as 30-50 million species of insects, making this pervasive terrestrial arthropod group, 97 per cent of global biodiversity.

Insects and their relatives are the most dominant and important group of terrestrial organisms besides humans, that affect life on Earth. They affect human life in a multitude of ways, both for good and bad. Profound ignorance about insect life permeates most of human society, even among the highly educated people. Insects and their relatives, in fact, are little credited for their beneficial environmental services but overblamed for their destructive activities. Despite lack of general human interest in insects, R.O. Wilson, wrote that they are "the little things that run the world".

Insects and their relatives live on all continents and occupy microhabitats from deep in the soil and underground aquifers to the top of trees and mountains, among the feathers of penguins on Antartica, and even deep into caves and in our eyebrows. Many lineages have evolved adaptations for living on and under ice fields, others at the margins of hot springs, and still others on the open ocean, land arthropods, by virtue of their pervasiveness, are incredibly important to the balance of life within ecosystems. Insects and their relatives eat virtually everything and compete even for the rocks under which they hide, mate, and rear their young ones. For one thing, if there is no insect species most broadleaf trees and shrubs would not be pollinated, and there would be no fruits and seeds. For another, instead of penetrating dead matter, decomposers such as bacteria and fungi would live only on the surface, taking years or perhaps millennia to break the substrate down into recyclable nutrients for plants, and thus soils would be much less fertile. Many fish and birds, and even some mammals, would have no food and would cease to exist.

Among the insects, the bettles are the most specious, the most pervasive, and the most widespread across the face of the globe. During dry seasons in tropical forests, they are the third most numerous individuals, after ants and termites, making up 12 per cent of the total insects community after ants and termites.

Beetles are found everywhere on our planet except in the deep sea. However, they do occur commonly in the sea's intertidal zone and estuarine salt flats. Beetles even occurred on Antarctica not long ago. Most families of beetles, about 140 of them, are worldwide in distribution, and their species provide equivalent ecological services wherever they occur. The "play" is generally the same everywhere, only the "actors" themselves change from place to place.

The history shows that two major faunal changes took place, the first in the Mid Jurassic Period, when primitive lineages of beetles lost their dominance, and the second in the mid-Cretaceous Period, when modern forms acquired dominance over all other terrestrial arthropods. In terms of species and number of guilds, they still have this dominance in nearly every biotype. By any broad measure, beetles are the most successful lineage of complex organisms ever to have evolved.

The described species of beetles, about 400,000 above comprise about 25 per cent, of all described species on Earth. This dominance of beetle taxa in the literature has resulted in Coleoptera being perceived as Earth's most specious taxon. Thus, it has garnered further taxonomic attention from young taxonomists which in turn has resulted in more species of beetles being described than in other groups. Beetles are relatively easy to collect, prepare, and describe, significantly adding to their popularity. Such unevenness in taxonomic efforts may or may not give us a false picture of true relative insect diversities. Nevertheless, the dominance of beetles has been used to arrive at an estimate of 3" million insects overall, and even to designate the group most endeared to God. While this dominance may be arguable either scientifically or philosophically, it is certainly interesting.

The reasons probably lie in the overwhelming numbers of species, individuals, and the ever-plodding course of traditional taxonomy. Potential users of data on beetles simply have to wait too long to get names. Taxonomists have to wait too long to receive money to visit museums in which name-bearing type specimens are held; monographers take too long to produce documents with which users might identify their specimens by themselves and specialists are reluctant to take on a large identification load for other scientists, such as ecologists and conservation biologists.

Neotropical beetles are second only to ants and flies in numbers of free-ranging individuals of arthropods in the canopies and subcanopies of neotropical trees; Psocoptera are a distant third. However, per species, beetles are not abundant. Beetles participate in virtually all aspects of ecosystem processes; they are predators, herbivores, folivores, detritivores, scavengers, fungivores, woodeaters, and grazers, and they tunnel, mine and chew nearly every substrate. Some are ectoparasites, others are nest parasites, some even live in the fur of vertebrates. Still others are subsocial, with adults participating in the raising of young. A global perspective based on beetles could provide a much more finegrained view of biodiversity than the coarse grained one which we get from less specious groups such as jaguars, birds and monkeys, which heretofore have garnered most of the attention.

Collection of Data

The interface between insects and their environment is at a small resolution, the information they provide may well be critical for ecological restoration. Management will depend on what we really know rather than what we surmise. Conservation cannot now deal with insect information, but will be compelled to do so in the not-too-distant future. We will need a system for data gathering that is just now becoming available.

One goal of Janzen's ATBI is to inventory all the taxa within a given area. A biotic inventory includes finding the area's species, classifying them, making voucher collections, and storing these data in a way that they are easily retrievable. Additional information about the species, either gathered during the process of inventorying or added later from literature or follow-up studies, can be piled on top of the four basic elements in a growing database.

One goal of Janzen's ATBI is to inventory all the taxa within a given area. A biotic inventory includes finding the area's species, classifying them, making voucher collections, and storing these data in a way that they are easily retrievable. Additional information about the species, either gathered during the process of inventorying or added later from literature or follow-up studies, can be piled on top of the four basic elements in a growing database.

The first ATBI is destined to be at the Guanacaste Conservation Area (GCA), Costa Rica, a site with dry forest in lower elevations ascending through cloud forest and continuing intermediaries between these levels, based on the experience in nearby Panama with a similar range of habitats, it is estimated that GCA should have about 50,000 species of beetles. Since, these estimation can be only a first approximation, it is used for purposes of designing a sampling regime for the project; budget and time must be considered to be modifiable as the project narrows to better estimates.

Each species of tree, shrub, and herb/grass may have its own host-specific species of beetles. Riparian strands in various watersheds will have different types of substrates, water quality, vegetation, etc., contributing to their distinctive biodiversities. In addition, the GCA is distinctly seasonal, hence both dry and wet seasons need to be sampled for each microhabitat.

The sampling regime must consider the above in its attempt to record as many species as possible in the shortest amount of time. The guiding principles are as follows:

Phase 1 : mass co-occurrence sampling; rapid processing with bulk cold storage; identification process using matching specimens; interim naming with alphanumerics; accumulation of data using linked spread sheets, including curves showing sampling progress and character filing with the Quick Taxonomic Assessment System (QTES).

Phase 2 : send target taxa and QTES data into the taxasphere for formal species names;

Phase 3: replace EXCEL 4.0 spreadsheet and QTES interim names with formal one, transfer these data to the database at Instituto Nacional de Biodiversidad (INBio).

Phase 4: generate illustrations and three-dimensional laser images; produce documents.

An Agenda for Sampling Beetles in an ATBI Sampling

The following criteria must be met for acquiring samples of beetles that can provide a reasonable inventory and serve both immediate and future needs of research, as well as determine to an order of magnitude the species present in the target area for us in subsequent sampling projects:

1. Sampling assumes the use of a fogger and three per cent Resmethrin for all microhabitats from one in above ground through the canopy rim; this should capture one to seven species per inch of foiage and much more in compacted microhabitats such as suspended dry leaves, vine tangles, and complex canopies. Leaf, litter and soil layers are sampled by photoeclectros and sifting/Toplgren extractor techniques. Berlese banks can substitute for Tolgren, if electricity is available. The stratum of herbs and grasses is sweep-sampled by sweep-netting.

Methods of trapping by attraction and even passive traps that catch flying insects produce catch without biocontoxt, i.e., specimens that are not tied to any microhabitat, substrate, host plant, etc. Much time and effort goes into preparing, identifying, and storing such bulk lots, yet the quality of data is at the lowest level. These methods of collecting are simply not worth the efforts, unless one is interested solely in recording the presence of species in the general area or in building collections. However, these techniques can be used as a test of the methods that incorporate biocontext to determine if microhabitats exist that are not being sampled with the other techniques.

(2) A standard set of field data includes precise locality; type of forest; type of microhabitat and its volume or surface area; information on species of plants; date;

and collector(s). Lot numbers are assigned to each individual fogging collection, sifting series, photoeclector sample, sweep series, etc. Thus, all specimens taken from the same microhabitat or plant or trap at the same time get the same lot number so that the set of species, including non-beetles, can be reassembled at a later date if desired. Non-beetle specimens will be directed to another Taxonomic Working Group (TWG), along with appropriate sets of data.

(3) Specimens of beetles are preserved in 70 per cent alcohol in the field. Alcohol must be changed the same day at the lab and subsequently each time the specimens undergo processing. Egg carton inserts are placed in funnels or on suspended sheets to catch dry specimens. After these are selected from the carton surface, the remaining specimens are dumped into the alcohol bottle.

(4) Sampling design involves taking replicate microhabitat samples in sets of 10 throughout each type of forest and open area. During preparation and data entry of the 10 samples, species accumulation curves and Chao's estimator track the progress of the inventory. A complete inventory for smaller families will require fewer replicates, but the leaf-beetles and weevils will require more than 10 replicates, based on data from over 5,000 species acquired at Pakitza, Peru, from 1988-1992. A decision needs to be made at the outset as to when to stop, because it will not be "humanly" possible to get the "last" species on the list in the larger families, however, 100 per cent likely will be reached in smaller families.

Preparation

All tropical forest samples are replete with beetles. The object of preparation should be to make the species and their whereabouts and abundance known in the shortest amount of time possible. Traditional preparation of all collected specimens, therefore, is not feasible. The following method leads to one prepared specimen per species per samples, with cold storage of the bulk lots that can easily be accessed later by taxonomists and other workers who need series.

(1) Each sample lot is sorted to families using a 6.2 cm white ceramic dish with 70 per cent alcohol. Families are gathered in small plastic lids set inside the bottom of a petri dish, the top of which is ringed with vaseline to create a seal when the specimens are sitting unworked. Parataxonomists and beginning graduate students can be trained to sort at this level quickly.

(2) Each family then is sorted in sequence to species, with one good specimen selected for pinning/pointing. The specimen is placed on damp filter paper inside another petri dish. On the filter paper, numbers 1 to 20 are written and the chosen specimen is placed next to a number according to its abundance in the sample. If over 20 specimens of a species are in a sample, a small label is written with the number, and this label is affixed to the pin to be removed later in the process. The rest of the counted but unprepared specimens are returned to the lot vial of 70 per cent alcohol, bar-coded, and sent to cold storage. Sorting to species across the Coleoptera can be done only by a highly trained taxonomist with long experience, and this person becomes the key to the entire project. Preparation and storage procedures can be handled by a technically trained person.

(3) Each specimen from the filter paper is pinned or pointed with Elmer's glue to pins or pre-prepared points in the traditional manner and placed in a unit, tray with strips of numbers sequenced from 1 to 20. Each specimen is aligned next to one of these numbers according to its abundance in the sample.

(4) Preprinted labels are attached to each specimen as it is "identified by matching" using the synoptic collection. The name of the species is an alphanumeric in the form of "family code + number" and lot number. All families of Coleoptera have a standard code of four letters. The margin of the label is colour-coded with pencil for instant recognition of microhabitat, although the lot number references this too. Once a family is represented by more than 100 species, "identifying-by-matching" becomes less and less efficient. For very large families, such as weevils, staphylinids, and chrysomelids, use of QTES is recommended.

Interim identification

Each prepared specimen from a sample is compared with its corresponding family-level synoptic collection. Smaller families are easy to keep in one or a few units. Larger families may be subdivided by subfamily so that the amount of matching necessary for recognizing the specimen's status is kept to a minimum. As a specimen is identified or is kept to a minimum or recognized as a species new to the synoptic collection, it is placed either in an interim unit tray awaiting entry of the data before going to the duplicate collection or added to the synoptic collection, from where its data are entered. All species of a family that are sorted from the sample are labelled, then the data are entered in EXCEL before preparing the next family.

Data storage

EXCEL linked spreadsheet templates for families of beetles contain about 13 kilobytes of forms that are based on microhabitats. Entry of data from a sample involves simply number of specimens per species per lot. The programme automatically computes all basic information and accumulates the data on summary sheets for easy viewing. The programme is exceedingly user-friendly.

Building Collections

The resulting synoptic-unit trays of families of beetles are ready for special scrutiny at any time during the process if the specialists is on-site to make formal identifications. The duplicate collection built from second through occurrences of a species across samples can be sent through the taxasphere regularly and results can be fed back into the EXCEL data system, making it easy to move the information to the INBio standard data files. As additional microhabitat replicates are sampled and specimens processed. Those species represented by uniques in the synoptic collection will be duplicated and can then be sent through the taxasphere..

Serious taxonomists who must do a revolutionary study immediately can read the database to find lots with series and arrange to extract those from cold storage themselves. Common species that are found in many or most lots will have that many more prepared specimens ready for study in the duplicate collection.

15

BIOLOGICAL DIVERSITY CONVENTION

Biological diversity is being significantly reduced by certain human activities. Aware of the general lack of information and knowledge regarding biological diversity and of the urgent need to develop scientific, technical and institutional capacities to provide the basic understanding upon which to plan and implement appropriate measures.

Noting further that ex situ measures, preferably in the country of origin, also have an important role to play. Recognising the close and traditional dependence of any indigenous and local communities embodying traditional lifestyles on biological resources, and the desirability of sharing equitably benefits arising from the use of traditional knowledge, innovations and practices relevant to the conservation of biological diversity and the sustainable use of its components. Recognising also the vital role that women play in the conservation and sustainable use of biological diversity and affirming the need for the full participation of women at all levels of policy-making and implementation for biological diversity conservation.

Stressing the importance of, and the need to promote, international, regional and global cooperation among States and intergovernmental organisations and the non-governmental sector for the conservation of biological diversity and the sustainable use of its components. Acknowledging further that special provision of new and additional financial resources and appropriate access to relevant technologies can be expected to make a substantial difference in the world's ability to address the loss of biological diversity.

Noting in this regard the special conditions of the least developed countries and small island States. Acknowledging that substantial investments are required to conserve biological diversity and that there is the expectation of a broad range of environmental, economic and social benefits from those investments. Recognising that economic and social development and poverty eradication are the first and overriding priorities of developing countries.

Aware that conservation and sustainable use of biological diversity is of critical importance for promoting the food, health and other needs of the growing world population, for which purpose access to and sharing of both genetic resources and technologies are essential. Noting that, ultimately, the conservation and sustainable use of biological diversity will strengthen friendly relations among States and contribute to peace for humankind.

Desiring to enhance and complement existing international arrangement for the conservation of biological diversity and sustainable use of its components, and determined to conserve at sustainable use, biological diversity for the benefit of present and future generations have been agreed as follows :

Article 1: Objectives

The objectives of this Convention, to be pursued in accordance with its relevant provisions, are the conservation of biological diversity, the sustainable use of its components and the fair and equitable sharing of the benefits arising out of the utilisation of genetic resources, including by appropriate access to genetic resources and by appropriate transfer of relevant technologies, taking into account all rights over those resources and to technologies, and by appropriate funding.

Article 2: Use of Terms

For the purposes of this Convention:

"Biological diversity" means the variability among living organisms from all sources, including, inter alia, terrestrial, marine and other aquatic ecosystems and the ecological complexes of which they are part: this includes diversity within species, between species and of ecosystems.

"Biological resources" includes genetic resources, organisms or parts thereof, populations, or any other biotic component of ecosystems with actual or potential use or value for humanity. "Biotechnology" means any technological application that uses biological systems, living organisms, so derivatives therefore, to make or modify products or process for specific use.

"Country of origin of genetic resources" means the country which possesses those genetic resources is in situ conditions. "Country providing genetic resources" means the country supplying genetic resources collected from in situ sources, including populations of both wild and domesticated species, or taken from ex situ sources, which may or may not have originated in that country "Domesticated or cultivated species" means species in which the evolutionary process has been influenced by humans to meet their needs.

"Ecosystem" means a dynamic complex of plant, animal and microorganism communities and their non-living environment interacting as a functional unit.

"Ex situ conservation" means the conservation of components of biological diversity outside their natural habitats.

"Genetic Material" means any material of plant, animal, microbial or other origin containing functional units of heredity.

"Genetic resources" means genetic material of actual or potential value.

"Habitat" means the place or type of site where an organism or population naturally occurs.

"In situ conditions" means conditions where genetic resources exist within ecosystems and natural habitats and the maintenance and recovery of viable populations of species in their natural surrounding and, in the case, of domesticated or cultivated species, in the surrounding where they have developed their distinctive properties.

"Protected area" means a geographically defined area which is designated or regulated and managed to achieve specific conservation objectives.

"Regional economic integration organisation" means an organisation constituted by sovereign States of a given region, to which its member States have transferred competence in respect of matters governed by this Convention and which has been duly authorised, an accordance with its internal procedures, to sign, ratify, accept, approve or accede to it.

"Sustainable use" means the use of components of biological diversity in a way and at a rate that does not lead to long term decline of biological diversity, thereby maintaining its potential to meet the needs and aspirations of present and future generations.

"Technology" includes biotechnology.

Article 3: Principle

States have, in accordance with the Charter of the United Nations and the principles of international law, the sovereign right to exploit their own resources pursuant to their own environmental policies, and the responsibility to ensure that activities within their jurisdiction or control for not causing any damage to the Environment of other States or of areas beyond the limits of national Jurisdiction.

Article 4: Jurisdictional Scope

Subject to the rights of other States, and except as otherwise expressly provided in this Convention, the provisions of this Convention apply, in relation to each Contracting Party :

(a) In the case of components of biological diversity, in areas within the limits of its national jurisdiction; and

(b) In the case of processes and activities, regardless of where their effects occur, carried out under its jurisdiction or control, within the area of its national jurisdiction or beyond the limits of national jurisdiction.

Article 5: Cooperation

Each Contracting Party shall, as far as possible and as appropriate, cooperate with other contracting parties, directly or, where appropriate, through competent international organisations; in respect of areas beyond national jurisdiction and on other matters of mutual interest, for the conservation and sustainable use of biological diversity.

Article 6: General Measures for Conservation and Sustainable Use

Each Contracting Party shall, in accordance with its particular conditions and capabilities :

(a) Develop national strategies, plans or programmes for the conservation and sustainable use of biological diversity or adapt for this purpose existing strategies, plans or programmes which shall reflect, inter alia, the measures set out in this Convention relevant to the Contracting Party concerned; and

(b) Integrate as far as possible and as appropriate, the conservation and sustainable use of biological diversity into relevant sectoral or cross-sectoral plans, programmes and policies.

Article 7: Identification and Monitoring

Each Contracting Party shall, as far as possible and as appropriate, in particular for the purposes of Articles 8 to 10.

(a) Identify components of biological diversity important for its conservation and sustainable use having regard to the indicative list of categories set down in Annex. I;

(b) Monitor, through sampling and other techniques, the components of biological diversity identified pursuant to subparagraph (a) above, paying particular attention to those requiring urgent conservation measures and those which offer the greatest potential for sustainable use;

Identify process and categories of activities which have or are likely to have significant adverse impacts on the conservation and sustainable use of biological diversity, and monitor their effects through sampling and other techniques; and maintain and organise by any mechanism data derived from identification and monitoring activities pursuant to subparagraphs (a), (b) and (c) above.

Article 8: In situ Conservation

Each Contracting Party shall, as far as possible and as appropriate:

(a) Establish a system of protected areas or areas where special measures need to be taken to conserve biological diversity;

(b) Develop, where necessary guidelines for the selection, establishment and management of protected areas where special measures need to be taken to conserve biological diversity;

(c) Regulate or manage biological resources for the conservation of biological diversity within or outside protected areas with a view to ensuring their conservation and sustainable use;

(d) Promote the protection of ecosystems, natural habitats and maintenance of viable populations of species in natural surroundings;

(e) Promote environmentally sound and sustainable development in areas adjacent to protected areas with a view to furthering protection of these areas;

(f) Rehabilitate and restore degraded ecosystems and promote the recovery of threatened species, inter alia, through the development and implementation of plans or other management strategies;

(g) Establish or maintain means to regulate, manage or control the risks associated with the use and release of living modified organisms resulting from biotechnology which are likely to have adverse environmental impacts that could affect the conservation and sustainable use of biological diversity, taking also into account the risks to human health;

(h) Prevent the introduction of control or eradicate those alien species which threaten ecosystems, habitats or species;

(i) Endeavour to provide the conditions needed or compatibility between present uses and the conservation of biological diversity and the sustainable use of its components;

(j) Subject to its national legislation, respect, preserve and maintain knowledge, innovations and practices of indigenous and local communities embodying traditional lifestyles relevant for the conservation and sustainable use of biological diversity and promote their wider application with the approval and involvement of the holders of such knowledge, innovations and practices and encourage the equitable sharing of the benefits arising from the utilisation of such knowledge, innovations and practices;

(k) Develop or maintain necessary legislation and/or other regulatory provisions for the protection of threatened species and populations;

(1) Where a significant adverse effect on biological diversity has been determined pursuant to Article

7, regulate or manage the relevant processes and categories of activities; and

(m) Cooperate in providing financial and other support for in situ conservation outlined in subparagraphs (a) to (1) above, particularly to developing countries.

Article 9: Ex Situ Conservation

Each Contracting Party shall, as far as possible and as appropriate, and predominantly for the purpose of complementing in situ measures.

(a) Components of biological diversity, of such components;

(b) Conservation of and research on plants, animals and microorganisms;

(c) Adopt measures for the recovery and rehabilitation of threatened species and for their reintroduction into appropriate conditions;

(d) Regulate and manage collection of biological resources from purposes so as not to threaten ecosystem and in situ population of species except where special temporary ex situ measures are required under subparagraph (c) above; and

(e) Cooperate in providing for ex situ conservation outlined in subparagraphs (a) to (d) above and in the establishment and maintenance of ex situ conservation facilities in developing countries.

Article 10: Sustainable use of components of biological diversity

Each Contracting Party shall, as far as possible and as appropriate

(a) Integrate consideration of the conservation and sustainable use of biological resources;

(b) Adopt measures relating to the use of biological resources to avoid or minimise adverse impacts on biological diversity;

(c) Protect and encourage customary use of biological resources to avoid or minimise adverse impacts on biological diversity;

(d) Support local populations to develop and implement remedial action in degraded areas where biological diversity has been reduced; and

(e) Encourage cooperation between its governmental authorities and its private sector in developing methods for sustainable use of biological resources.

Article 11: Incentive Measures

Each Contracting Party shall, as far as possible and as appropriate, adopt economically and socially sound measures that act as incentives for the conservation and sustainable use of components of biological diversity.

Article 12: Research and Training

The Contracting Parties, taking into account the special needs of developing countries, shall:

(a) Establish and maintain programmes for scientific and technical education and training in measures for the identification, conservation and sustainable use of biological diversity and its components and provide support for such education and training for the specific needs of developing countries;

(b) Promote and encourage research which contributes to the conservation and sustainable use of biological diversity, particularly in developing countries inter alia, in accordance with decisions of the Conference of the Parties taken consequence of recommendations of the subsidiary Body on Scientific and Technical Advice; and

(c) In keeping with the provisions of Articles 10, 18 and 20, promote and cooperate in the use of scientific advances in biological diversity research in developing methods for conservation and sustainable use of biological resources.

Article 13: Public Education and Awareness

The Contracting Parties shall:

(a) Promote and encourage understanding of the importance of, and the measures required for the conservation of biological diversity, as well as its propagation through media, and the inclusion of these topics in educational programmes; and

(b) Cooperate, as appropriate, with other States and international organisations in developing educational and public awareness programmes, with respect to conservation and sustainable use of biological diversity.

Article 14: Impact Assessment and Minimising Adverse Impacts

1. Each Contracting Party, as far as possible and as appropriate, shall:

(a) Introduce appropriate procedures requiring environmental impact assessment of its proposed projects that are likely to have significant adverse effects on biological diversity with a view to avoiding or minimising such effects and, where appropriate, allow for public participation in such procedures.

(b) Introduce appropriate arrangements to ensure that the environmental consequence, of its programmes and policies that are likely to have significant adverse impacts on biological diversity are duly taken into account;

(c) Promote, on the basis of reciprocity, notification, exchange of information and consultation on activities under their jurisdiction or control which are likely significantly to affect adversely the biological diversity of other States or areas beyond the limits of national jurisdiction, by encouraging the conclusion of bilateral, regional or multilateral arrangements as appropriate;

(d) In the case of imminent or grave danger or damage originating under its jurisdiction or control to biological diversity within the area under jurisdiction of other States or in areas beyond the limits of national jurisdiction of other States or in areas beyond the limits of national jurisdiction, notify immediately the potentially affected States of such danger or damage, as well as initiate action to prevent or minimise such danger or damage;

(e) Promote national arrangements for emergency responses to activities or events, where caused naturally or otherwise, which present a grave and imminent danger to biological diversity and encourage international cooperation to supplement such national efforts and, where appropriate and agreed by the States or regional economic integration organisations concerned, to establish joint contingency plans.

2. The Conference of the Parties shall examine, on the basis of studies to be carried out, the issue of liability and redress, including restoration and compensation, for damage to biological diversity, except where such liability is a purely internal matter.

Article 15: Access to Genetic Resources

1. Recognising the sovereign rights of States over their natural resources, the authority to determine access to genetic resources rests with the natural governments and is subject to national legislation.

2. Each Contracting Party shall endeavour to create conditions to facilitate access to genetic resources for environmentally sound uses by other Contracting Parties and to impose restrictions that run counter to the objectives of this Convention.

3. For the purpose of this Convention, the genetic resources being provided by a Contracting Party, as referred to in this Article and Article 16 and 19,

origin of such resources or by the Parties that have acquired the genetic resources in complete with this Convention.

4. Access, where granted, shall be on mutually agreed terms and subject to the provisions of the Article.

5. Access to genetic resources shall be subject to prior informed consent of the Contracting Party providing such resources, unless otherwise determined by that Party.

6. Each Contracting Party shall endeavour to develop and carry out scientific research based on genetic resources provided by other Contracting Parties with the full participation of, and where possible in, such Contracting Parties.

7. Each Contracting Party shall take legislative, administrative or policy measures, as appropriate, and in accordance with Articles 16 and 19 and where necessary through the financial mechanism established by Articles 20 and 21, with the aim of starting in a fair and equitable way the results of research and development and the benefits arising from the commercial and other utilisation of genetic resources with the Contracting Party providing such resources. Such sharing shall be upon mutually agreed terms.

Article 16: Access to and Transfer of Technology

1. Each Contracting Party, recognising that technology includes biotechnology, and that both access to and transfer of technology among Contracting Parties are essential elements for the attainment of the objectives of this Convention, undertakes subjects to the provisions of this Article to provide and/or facilitate access for and transfer to other Contracting Parties of technologies that are relevant to the conservation and sustainable use of biological

diversity or make use of genetic resources and do not cause significant damage to the environment.

2. Access to and transfer of technology referred to in paragraph 1 above to developing countries shall be provided and/or facilitated under fair and most favourable terms, including on concessional and preferential terms where agreed, and where necessary in accordance with the financial mechanism established by Articles 20 and 21. In the case of technology to patents and other intellectual property rights, such access and transfer shall be provided on terms with recognise and are consistent with the adequate and effective protection of intellectual property rights. The application of this paragraph shall be consistent with paragraphs 3, 4 and 5 below.

3. Each Contracting Party shall take legislative, administrative or policy measures, as appropriate, with the aim that Contracting Parties, in particular those that are developing countries, which provide genetic resources are provided access to and transfer of technology which makes use of those resources... on mutually agreed terms, including technology protected by patents and other intellectual property rights, where necessary through the provisions of Articles 20 and 21 and in accordance with international law and consistent with paragraphs 4 and 5 below.

4. Each Contracting Party shall take legislative, administrative or policy measures, as appropriate, with the aim that the private sector facilitates access to, joint development and transfer of technology referred to in paragraph 1 above for the benefit of both governmental institutions and the private sector of developing countries and in this regard shall abide by the obligations included in paragraphs 1, 2 and 3 above.

5. The Contracting Parties, recognising that patents and other intellectual property rights may have an influence on the implementation of this Convention, shall cooperate in this regard subject to national legislation and international law in order to ensure that such rights are supportive of and do not run counter to its objectives.

Article 17: Exchange of Information

1. The Contracting Parties shall facilitate the exchange of information from all publicly available source, relevant to the conservation and sustainable use of biological diversity, taking into account the special needs of developing countries.

2. Such exchange of information shall include exchange of results of technical, scientific and socio-economic research, as well as information on training and surveying programmes, specialised knowledge, indigenous and traditional knowledge as such and in combination with the technologies referred to in Article 16, paragraph 1. It shall also, where feasible, include repatriation of information.

Article 18: Technical and Scientific Cooperation

1. The Contracting Parties shall promote international technical and scientific cooperation in the field of conservation and sustainable use of biological diversity, where necessary through the appropriate international and national institutions.

2. Each Contracting Party shall promote technical in particular developing countries, in implementing this Convention, inter alia, through the development and implementation of national policies. In promoting such cooperation, special attention should be given to the development and strengthening of national capabilities by means of human resources development and institution building.

3. The Conference of the Parties, at its first meeting, shall determine how to establish a clearing house mechanism to promote and facilitate technical and scientific cooperation.

4. The Contracting Parties shall, in accordance with national legislation and policies, encourage and develop methods of cooperation for the development and use of technologies, including indigenous and traditional technologies, in pursuance of the objectives of this Convention. For this purpose, the Contracting Parties shall also promote cooperation in the training of personnel and exchange of experts.

5. The Contracting Parties shall, subject to mutual agreement, promote the establishment of joint research programmes and joint ventures for the development of technologies relevant to the objectives of this Convention.

Article 19: Handling of Biotechnology and Distribution of Its Benefits

1. Each Contracting Party shall take legislative, administrative or policy measures, as appropriate, to provide for the effective participation in biotechnological research activities by those Contracting Parties, especially developing countries, which provide the genetic resources for such research, and where feasible in such Contracting Parties.

2. Each Contracting Party shall take all practicable measures to promote and advance priority access on a fair and equitable basis by Contracting Parties, especially developing countries, to the results and benefits arising from biotechnologies based upon genetic resources provided by those Contracting Parties. Such access shall be on mutually agreed terms.

3. The Parties shall consider the need for and modalities of a protocol setting out appropriate procedures, including, in particular advance informed agreement, in the field of the safe transfer, handling and use of any living modified organism resulting from biotechnology that may have adverse effect on the conservation and sustainable use of biological diversity.

4. Each Contracting Party shall, directly or by requiring any natural or legal person under its jurisdiction providing the organisms referred to in paragraph 3 above, provide any available information about the use and safety regulations required by that Contracting Party in handling such organisms as well as any available information on the potential adverse impact of the specific organisms concerned to the Contracting Party into which those organisms are to be introduced.

Article 20: Financial Resources

1. Each Contracting Party undertakes to provide, in accordance with its capabilities, financial support and incentives in respect of those national activities which are intended to achieve the objectives of this Convention, in accordance with its national plans, priorities and programmes.

2. The developed country Parties shall provide new and additional financial resources to enable developing country Parties to meet the agreed full incremental costs to them of implementing measures which fulfil the obligations of this Convention and to benefit from its provisions and which costs are agreed between a developing country Party by the institutional structure, referred to in Article 21, in accordance with policy, strategy, programme priorities and eligibility criteria and an indication list of incremental costs established by the

Conference of the Parties. Other Parties, including countries undergoing the process of transition to a market economy, may voluntarily assume the obligations of the developed country Parties. For the purpose of this Article, the Conference of the Parties shall at its first meeting establish a list of developed country Parties and other Parties which voluntarily assume the obligations of the developed country Parties.

The Conference of the Parties shall periodically review and if necessary, amend the list. Contributions from other countries and source on a voluntary basis would also be encouraged. The implementation of these commitments shall take into account the need for adequacy, predictability and timely flow of funds and the importance of burden-sharing among the contributing Parties included in the list.

3. The developed country Parties may also provide, and developing country Parties avail themselves of, financial resources related to the implementation of this Convention through bilateral, regional and other multilateral channels.

4. The extent to which developing country Parties will effectively implement their commitments under the Convention will depend on the effective implementation by developed country Parties of their commitments under the Convention related to financial resources and transfer of technology and will take fully into account the fact that economic and social development and eradication of poverty are the first and overriding priorities of the developing country Parties.

5. The Parties shall take full account of the specific needs and special situation of least developed countries in their actions with regard to funding and transfer of technology.

6. The Contracting Parties shall also take into consideration the special conditions resulting from the dependence on distribution and location of biological diversity within developing country Parties, in particular small island States.

7. Consideration shall also be given to the special situation of developing countries, including those that are most environmentally vulnerable, such as those with arid and semi-arid zones, coastal and mountainous areas.

Article 21: Financial Mechanism

1. There shall be a mechanism for the provision of financial resources to developing country Parties for purposes of this Convention on a grant or concessional basis, the essential elements of which are described in this Article. The mechanism shall function under the authority and guidance of, and be accountable to, the Conference of the Parties for purposes of this Convention. The operations of mechanism shall be carried out by such institutional structure as may be decided upon by the Conference of the Parties at its first meeting. For purposes of this Convention, the Conference of the Parties shall determine the policy, strategy, programme priorities and eligibility criteria relating to the access to and utilisation of such resources. The contributions shall be such as to take into account the need for predictability, adequacy and timely flow of funds referred to in Article 20 in accordance with the amount of resources needed to be decided periodically by the Conference of the Parties included in the list referred to in Article 20, paragraph 2. Voluntary contributions may also be made by the developed country parties and by other countries and sources. The mechanism shall operate within a democratic and transparent system of governance.

2. Pursuant to the objectives of this Convention, the Conference of the Parties shall at its first meeting determine the policy, strategy and programme priorities, as well as detailed criteria and guidelines for eligibility for access to and utilisation of the financial resources including monitoring and evaluation on a regular of such utilisation. The Conference of the Parties shall decide on the arrangements to give effect to paragraph 1 above after consultation with the institutional structure entrusted with the operation of the financial mechanism.

3. The Conference of the Parties shall review the effectiveness of the mechanism established under this Article, including the criteria and guidelines referred to in paragraph 2 above, not less than two years after the entry into force of this Convention and thereafter on a regular basis. Based on such review, it shall take appropriate action to improve the effectiveness of the mechanism if necessary.

4. The Contracting Parties shall consider strengthening existing financial institutions to provide financial resources for the conservation and sustainable use of biological diversity.

Article 22: Relationship with other International Conventions

1. The provisions of the present Convention shall not affect the rights and obligations of any Contracting Party deriving from any existing international agreement, except where the exercise of those rights and obligations would cause serious damage or a threat to biological diversity.

2. Contracting Parties shall implement this Convention with respect to the marine environment consistently with the rights and obligations of States under the law of the sea.

Article 23: Conference of the Parties

1. A Conference of the Parties is hereby established. The first meeting of the Conference of the Parties shall be convened by the Executive Director of the United Nations Environment Programme not later than one year after the entry into force of this Convention. Thereafter, ordinary meetings of the Conference of the Parties shall be held at regular intervals to be determined by the Conference at its first meeting.
2. Extraordinary meetings of the Conference of the Parties shall be held at such other times as may be deemed necessary by the Conference or at the written request of any Party, provided that within six months of the request being communicated to them by the Secretariat, it is supported by at least one-third of the Parties.
3. The Conference of the Parties shall by consensus agree upon and adopt rules of procedure for itself and for any subsidiary body it may establish, as well as financial rules governing the funding of the Secretariat. At each ordinary meeting, it shall adopt a budget for the financial period until the next ordinary meeting.
4. The Conference of the Parties shall keep under review the implementation of the present Convention, and, for this purpose, shall:
   (a) Establish the form and the intervals for transmitting the information to be submitted in accordance with Article 26 and consider such information as well as reports submitted by any subsidiary body;
   (b) Review scientific, technical and technological advice on biological diversity provided in accordance with Article 25;

(c) Consider and adopt, as required, protocols in accordance with Article 28.

(d) Consider and adopt, as required, in accordance with Articles 29 and 30, amendments to this Convention and its annexes;

(e) Consider amendments to any protocol, as well as to any annexes thereto, and, if so decided, recommend their adoption to the parties to the protocol concerned;

(f) Consider and adopt, as required, in accordance with Article 30, additional annexes to this Convention;

(g) Establish such subsidiary bodies, particularly to provide scientific and technical advice, as are deemed necessary for the implementation of this Convention;

(h) Contact, through the Secretariat, the executive bodies of conventions dealing with matters covered by this Convention with a view to establishing appropriate forms of cooperation with them; and

(i) Consider and undertake any additional action that may be required for the achievement of the purposes of this Convention in the light of experience gained in its operation.

5. The United Nations, its specialised agencies and the International Atomic Energy Agency, as well as any State not party to this Convention, may be represented to observers at meetings of the Conference of the Parties. Any other body or agency, whether governmental or non-governmental, qualified in fields relating to conservation and sustainable use of biological diversity which has informed the Secretariat of its wish to be represented as an observer at a meeting of the Conference of the Parties present object.

Article 24: Secretariat

1. A secretariat is hereby established. Its functions shall be:
   (a) To arrange for and service meetings of the Conference of the Parties provided for in Article 23;
   (b) To perform the functions assigned to it by any protocol;
   (c) To prepare report on the execution of its function under this Convention and present them to the Conference of the Parties;
   (d) To coordinate with other relevant international bodies, and in particular to enter into such administrative and contractual arrangements as may be required for the effective discharge of its functions; and
   (e) To perform such other functions as may be determined by the Conference of the Parties.
2. At its first ordinary meeting, the Conference of the Parties shall designate the secretariat from amongst those existing competent international organisations which have signified their willingness to carry out the secretariat functions under this Convention.

Article 25: Subsidiary body on scientific, technical and technological advice

1. A subsidiary body for the provision of scientific, technical and technological advice is hereby established to prove the Conference of the Parties and, as appropriate, its other subsidiary bodies with timely advice relating to the implementation of this Convention. This body shall be open to participation by all Parties and shall be multi-disciplinary. It shall comprise government representatives competent in the relevant field of expertise. It shall report regularly

to the Conference of the Parties on all aspects of its work.

2. Under the authority of and in accordance with guidelines laid down by the Conference of the parties, and upon its request, this body shall:

(a) Provide scientific and technical assessments of the status of biological diversity;

(b) Prepare scientific and technical assessments of the effects of types of measures taken in accordance with the provisions of this Convention.

(c) Identity innovative, efficient and state of art technologies and know-how relating to the conservation and sustainable use of biological diversity and advise on the ways and means of promoting development and/or transferring such technologies;

(d) Provide advice on scientific programmes and international cooperation in research and development related to conservation and sustainable use of biological diversity; and

(e) Respond to scientific, technical, technological and methodological questions that the Conference of the Parties and its subsidiary bodies may put to the body.

The functions terms of reference, organisation and operation of this body may be further elaborated by the conference of the parties.

Article 26: Reports

Each Contracting Party shall, at intervals to be determined by the Conference of the Parties, present to the Conference of the Parties reports on measures which it has taken for the implementation of the provisions of this Convention and their effectiveness in meeting the objectives of this Convention.

Article 27: Settlement of Disputes

1. In the event of a dispute between Contracting Parties concerning the interpretation or application of this Convention, the Parties concerned shall seek solution by negotiation.
2. If the parties concerned cannot reach agreement by negotiation, they may jointly seek the good offices of, or request mediation by a third party.
3. When ratifying, accepting, approving or acceding to this Convention, or at any time thereafter, a State or regional economic integration organisation may declare in writing to the Depositary that for a dispute not resolved in accordance with paragraph 1 or paragraph 2 above, it accepts one or both of the following means of dispute, settlement as compulsory;
4. If the parties to the dispute have not in accordance with paragraph 3 above, accepted the same or any procedure, the dispute shall be submitted to conciliation in accordance with Part 2 of Annex 11 unless the parties otherwise agree.
5. The provisions of this Article shall apply with respect to any protocol except as otherwise provided in the protocol concerned.

Article 28: Adoption of Protocols

1. The Contracting Parties shall cooperate in the formulation and adoption of protocols to this Convention.
2. Protocols shall be adopted at a meeting of the Conference of the Parties.
3. The text of any proposed protocol shall be communicated to the Contracting Parties by the Secretariat at least six months before such a meeting.

Article 29: Amendment of the Convention or Protocols

1. Amendments to this Convention may be proposed by any Contracting Party. Amendments to any protocol may be proposed by any Party to that protocol.
2. Amendments to this Convention shall be adopted at a meeting of the Conference of the Parties. Amendments to any protocol shall be adopted at meeting of the Parties to the Protocol in question. The text of any proposed amendment to this Convention or to any protocol, except as may otherwise be provided in such protocol, shall be communicated to the parties to the instrument in question by the secretariat at least six months before the meeting at which it is proposed, for adoption. The secretariat shall also communicate proposed amendments to the signatories to this Convention for information.
3. The Parties shall make every effort to reach agreement on any proposed amendment to this Convention or to any protocol by consensus. If all efforts at consensus have been exhausted, and no agreement reached, the amendment shall as a last resort be adopted by a two-thirds majority vote of the Parties to the instrument in question present and voting at the meeting, and shall be submitted by the Depositary to all Parties for ratification, acceptance or approval.
4. Ratification, acceptance or approval of amendments shall be notified to the Depositary in writing. Amendments adopted in accordance with paragraph 3 above shall enter into force among Parties having accepted them on the ninetieth day after the deposit of instruments of ratification, acceptance or approval by at least two-thirds of the Contracting Parties to this Convention or of the Parties to the protocol concerned, except as may otherwise be provided in

such protocol. Thereafter the amendments shall enter into force for any other Party on the ninetieth day after that Party deposits its instruments of ratification, acceptance or approval of the amendments.

5. For the purposes of the Article, "Parties present and voting" means parties present and casting an affirmative or negative vote.

Article 30: Adoption and Amendment of Annexes

1. The annexes to this Convention or to any protocol shall form an integral part of this Convention or of such protocol, as the case may be, and unless expressly provided otherwise, a reference to this Convention or its protocols constitutes at the same time a reference to any annexes thereto. Such annexes shall be restricted to procedural, scientific, technical and administrative matters.

2. Except as may be otherwise provided in any protocol with respect to its annexes, the following procedure shall apply to the proposal, adoption any entry into force of additional annexes to this Convention or of annexes to any protocol:

   (a) Annexes to this Convention or to any protocol shall be proposed and adopted according to the procedure laid down in Article 29.

   (b) Any Party that is unable to approve an additional annex to this Convention or an annex to any protocol to which it is party shall so notify the Depositary, in writing, within one year from the date of the communication of the adoption by the Depositary. The Depositary shall without delay notify all Parties of any such notification received. A Party may at any time withdraw a previous declaration of objection and the annexes shall thereupon enter into force for that Party subject to sub-paragraph (c) below:

(c) On the expiry of one year from the date of the communication of the adoption by the Depositary, the annex shall enter into force for all parties to this Convention or to any protocol concerned which have not submitted a notification in accordance with the provisions of sub-paragraph (b) above.

3. The proposal, adoption and entry into force of amendments to annexes to this Convention or to any protocol shall be subject to the same procedure as for the proposal, adoption and entry into force annexes to the Convention or annexes to any protocol.

4. If an additional annex or an amendment to an annex is related, an amendment to this Convention or to any protocol, the additional annex or amendment shall not enter into force until such time as the amendment to this Convention or to the protocol concerned enters into force.

Article 31: Right to Vote

1. Except as provided for in paragraph 2 below, each Contracting Party to this Convention or to any protocol shall have one vote.

2. Regional economic integration organisations, in matters within them competence, shall exercise their right to vote if their member States which are Contracting Parties to the Convention of the relevant protocol. Such organisations shall not exercise their right to vote in their member State exercise theirs, and vice versa.

Article 32: Relationship between this Convention and its Protocols

1. A State or a regional economic integration organisation may not become a party to a protocol unless it is, or becomes at the same time a Contracting Party to this Convention.

2. Decisions under any protocol shall be taken only by the Parties to the protocol concerned. Any Contracting Party that has not ratified accepted or approved a protocol may participate as an observer in any meeting of the parties to that protocol.

Article 33: Signature

This Convention shall be open for signature at Rio de Janeiro by all States and any regional economic integration organisation from the fifth day of June 1992 until.

Article 34: Ratification, Acceptance or Approval

1. This Convention and any protocol shall be subject to ratification acceptance or approval by States and by regional economic integration organisations. Instruments of ratification, acceptance or approval shall be deposited with the Depository.
2. Any organisation referred to in paragraph 1 above which becomes a Contracting Party to this Convention or any protocol without any or its member States being a Contracting Party shall be bound by all the obligations under the Convention or the protocol, as the case may be. In the case of such organisations, one or more of whose member States is a Contracting Party to the Convention or relevant protocol the organisation and its member States shall decide on their respective responsibilities for the performance of their obligations under the Convention or protocol, as the case may be in such cases, the organisation and the member States shall not be entitled to exercise rights under the Convention or relevant protocol concurrently.
3. In their instruments of ratification, acceptance or approval, the organisations, referred to in paragraph 1 above shall declare the extent of their competence with respect to the matters governed by the Convention or the relevant protocol. These organisations shall also inform the Depositary of any

relevant modification in the extent of their competence.

Article 35: Accession

1. This Convention and any protocol shall be open for accession by States and by regional economic integration organisations from the date on which the Convention or the protocol concerned is closed for signature. The instruments of accessions shall be deposited with the Depositary.
2. In their instruments of accession, the organisations referred to in paragraph 1 shall declare the extent of their competence with respect to the matters governed by the Convention or the relevant protocol. These organisations shall also inform the Depositary of any relevant modification in the extent of their competence.
3. The provisions of Article 34, paragraph 2, shall apply to regional economic integration organisations which accede to this Convention or any protocol.

Article 36: Entry into Force

1. This Convention shall enter into force on the ninetieth day after the date of deposit of the thirtieth instrument of ratification, acceptance approval or accession.
2. Any protocol shall enter into force on the ninetieth day after the date of deposit of the number of instruments of ratification acceptance, approval or accession, specified in that protocol, has been deposited.
3. For each Contracting Party which ratifies, accepts or approves this Convention or accedes thereto after the deposit of the thirtieth instrument of ratification, acceptance, approval or accession, it shall enter into force on the ninetieth day after the date of deposit

by such Contracting Party of its instrument of ratification, acceptance, approval or accession.

4. Any protocol, except as otherwise provided in such protocol, shall enter into force for a Contracting Party that ratifies, accepts or approves that protocol or accedes thereto after its entry into force pursuant to paragraph 2 above, on the ninetieth day after the date on which that Contracting Party deposits its instrument of ratification, acceptance, approval or accession, or on the date on which the Convention enters into force for that Contracting Party, whichever shall be the later.

5. For the purposes of paragraphs 1 and 2 above, any instrument deposited by a regional economic integration organisation shall not be counted as additional to those deposited by member States of such organisation.

Article 37: Reservations

No reservations may be made to this Convention.

Article 38: Withdrawals

1. At any time after two years from the date on which the present Convention has entered into force for a Contracting Party, that Contracting Party may withdraw from the Convention by giving written notification to the Depositary.

2. Any such withdrawal shall take place upon expiry of one Year after the date of its receipt by the Depositary, or on such later date as may be specified in the notification of the withdrawal.

3. Any Contracting Party which withdraws from this Convention shall be considered as also having withdraw from any protocol to which it is party.

Article 39: Financial Interim Arrangement

Provided that it has been fully restructured in accordance with the requirements of Article 21, the Global Environment

Facility of the United Nations Development Programme, the United Nations Environment Programme and the International Bank for Reconstruction and Development shall be the institutional structure referred to in Article 21 on an interim basis, for the period between the entry into force of this Convention and the first meeting of the Conference of the Parties or until the Conference of the Parties decides which institutional structure will be designated in accordance with Article 21.

Article 40: Secretariat interim arrangements

The secretariat to be provided by the Executive Director of the United Nations Environment Programme shall be the secretariat referred to in Article 24, paragraph 2, on an interim basis for the period between the entry into force of this Convention and the first meeting of the Conference of the Parties.

Article 41: Depository

The Secretary-General of the United Nations shall assume the functions of Depository of this Convention and any protocols.

Article 42: Authentic Texts

1. The original of this Convention, of which the Arabic, Chinese, English, French, Russian and Spanish texts are equally authentic, shall be deposited with the Secretary-General of the United Nations.
2. In witness whereof the undersigned, being duly authorised to that effect, have signed this Convention.
3. Done at Rio de Janeiro on this fifth day of June, one thousand nine hundred and ninety-two.

16

HOT SPOT INFORMATION

a. The Silent Valley: A Success Story

The process of evaluation of environmental impact assessment (EIA) was started for the first time in 1977 by the planning commission in India. Now it has become a routine procedure and its reports are not binding on the part of decision makers and the evaluation is also not compulsory of the part of Project owners/entrepreneurs. But before 1974 itself, it was found that a hydel project of Kerala Government was dropped due to the concern about its impact on environment and endangered species. The story of Silent Valley reveals that it was dropped only to provide due protection to the Lion-tailed macaque and some other endangered and endemic flowering plants of Western Ghats. The Kerala Forest Department documented a commemorative volume of the natural wonders of the silent valley and the campaigns that helped to preserve them. This chapter is prepared based on the article written by Anand Parthasarathy, appeared in Frontline, November 5, 1999.

Silent valley represents India's most precious environmental heritage and the visitors to this place, usually go back with a fund of fond memories. Naturalists viz., Vijayan and M. Balakrishnan who carried out the first systematic scientific survey of the Silent valley in 1977 said recently that they had less to fear from the forests fourlegged denizens the wild boar, the leopard, the panther and so on than from the employees of the Kerala State Electricity Board, who are posted to deal with intruders when there was considerable conflict between the board and environmentalists, who opposed the proposal to build a hydroelectric project in the Silent Valley. Suresh Flamon, the wildlife photographer trekked for weeks on end on the fringes of the valley trying to catch a glimpse of the rarest of a rare butterflies, the Travancore evening brown, which was last sighed in 1930 in the Periyar Tiger Reserve. The saga of Silent Valley is made up of such individual quests, which have helped to preserve this unique crucible of nature virtually intact.

Prompted by the widespread interest shown by conservationists during a seminar organised to make the first decennial of the Silent Valley National Park in 1995, the Kerala Forest Department brought out in the second week of October an anthology of 38 articles which together record the history of the silent valley the struggle by non-governmental organization in India and abroad to save from being overwhelmed by hydel project, its regeneration in a period of a decade as a protected national park together with a hard core authoritative documentation by naturalist and conservationists on the valleys unique flora and fauna. Silent Valley whispers of the vision is Edited by T.M. Manoharan, S. D. Biju, T. S. Mayor and P. S. Easa. The clinically exact documentation of every subspecies will be useful to naturalists, while the more personal reminiscences of those who were involved in the debate on the wildlife importance of the valley will provide the human touch for more casual readers.

Even with the major hydel project like Idukki already committed Kerala's northern districts faced a bleak power scenario in the late 1960s. The gorge of the Kunthipuzha as it tumbled down Western Ghats to Mannarkkad before joining to the Bharatpuza seemed an ideal location for a dam. The silent valley was a dense tableland of tropical evergreen forest, approximately 90 sq. km. in area and hemmed in on all sides by steep hills of the Nilgiri ranges (Fig. 32.1). The Kunthipuza traversed north to south descending 1500 meters in the process. It rained here almost throughout the year recording over 3000 mm of annual rainfall, with nearly 900 mm in July alone when the south-west monsoon is at its peak. Temperatures were always around 20°C approximately.

The British rulers of the Malabar notified it as a reserve in 1914. The suitability of the Valley for a hydel project was one of the kept secrets of British India, but survey parties penetrated the forest only in the 1950s. Finally in February 1973, the Planning Commission approved a 240 MW Silent Valley Hydro-electric Project, with an initial outlay of Rs. 25 Crores. Incidentally, this was the time when Mrs. Indira Gandhi was addressing the Stalkholm Conference about the concern for environment.

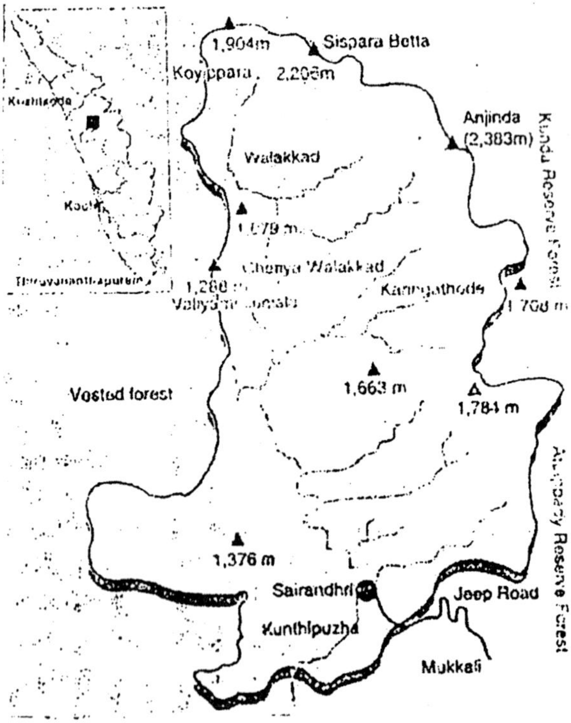

Fig. 32.1. Silent Valley National Park

But the government agencies were ignorant of the what Silent Valley held, individual naturalists, especially botanists and zoologists from colleges in northern Kerala, who had been trekking the valley for years in search of rare specimens were appalled at the developments. Around them grew the spontaneous moves to save the silent valley, M.K. Prasad a Botany lecturer in Kozikode as the convener of the Environmental brigade of the Kerala Shastra Sahitya Parishad (KSSP) brought the organization into campaign. Dr. S. Sathish Chantran Nair and Prof. M. P. Parameswaran carried the message across the state border to scientific societies in Chennai. Malayalam poet Sughath Kumari inspired by some of the KSSPs pamphlets, led a parallel movement among Keralas literati, which soon attracted well-known personalities of the state. As both Prasad and Sughat Kumari recalled in their contributions the Silent Valley protection campaign spread rapidly largely owing to the involvement of an academic literary combine.

Meanwhile nature lovers outside the state were networked. Dr. Salim Ali, Dr. Zafar Futehally from Mumbai, Dr. Madhav Gadgil and Dr. Parthasarathy from World Wildlife Fund for Nature, India, from Bangalore, Dr. Romulus Whitekar from Chennai joined this movement. They helped to add a national and international dimension to the issue flooded with telegrams from nature lovers and international organizations such as IUCN and WWF buoyed by a helpful on the spot study by the Principal Secretary in the Ministry of Agriculture, Dr. M. S. Swaminathan. Then Mrs. Indira Gandhi, the Prime Minister, took the political route-out of this potential development versus environmental crisis. She appointed Prof. M.G.K. Menon, Secretary, DST to carry out an appraisal of the issues involved.

Whispers of the Reason, is the book, includes the reminiscences of both Swaminathan and M.G.K. Menon and describes the fate of the valley following the swift of the political changes in New Delhi (succession of Prime Ministers) which came quickly, had to take decisions on the valley in less than two years. T.M. Manoharan main editor of the book, who is a Managing Director of Kerala Forest Development Corporation was divisional forest officer in-charge of the area that encompassed the valley during those troubled years. He was known to be sympathetic to all efforts at saving the valley.

Meanwhile, for saving the Silent Valley, campaigners took their struggle to the courts arguing the landmark suit filed by Joseph John on behalf of the Kochi-based, an NGO, Friends of Trees. Mrs. Indira Gandhi chose to read in the M.G.K. Menon report, which was presented in 1983, a clear message against the hydel scheme. In a swift order the state government was asked to call off the project and declare the valley as National Park. Mrs. Indira Gandhi, who took the crucial administrative decision, did not live to see the latter part of the order implemented. The National Park was formally inaugurated by Rajiv Gandhi in September 1985.

After the storm over, the Silent Valley had blown over a decade of intensive stock-taking was launched. Many institutions led by the Kerala Forest Research Institute (KFRI) got down to documenting the heritage of the valley in detail. In the process, they help to debunk many myths. Contrary to the belief that the Indian peninsula is poor in orchid wealth, with only a couple of dozen varieties, botanist K.S. Manilal says that the silent valley harbors have at least 108 varieties, many of them spectacularly large. In his article on orchids of the Tropical Botanical Garden and Research Institute (TBGRI) near Thiruvananthapuram lists every one of these species. Some, like the bicolor *Acanthephippium*, are found only in south India and Sri Lanka.

The Silent Valley is a repository of medicinal plants is brought out in the review by Sathish Kumar along with P. Pushpangadan, Director of the National Botanical Research Institute (NBRI) in Lucknow. Nearly 80 per cent of the drugs listed in standard pharmacopoeias and 66 per cent of the spices and aromatic plants used the world over are found in India. The Silent Valley is a valuable source of some important genetic variants. At least 21 flowering plants discovered in the valley

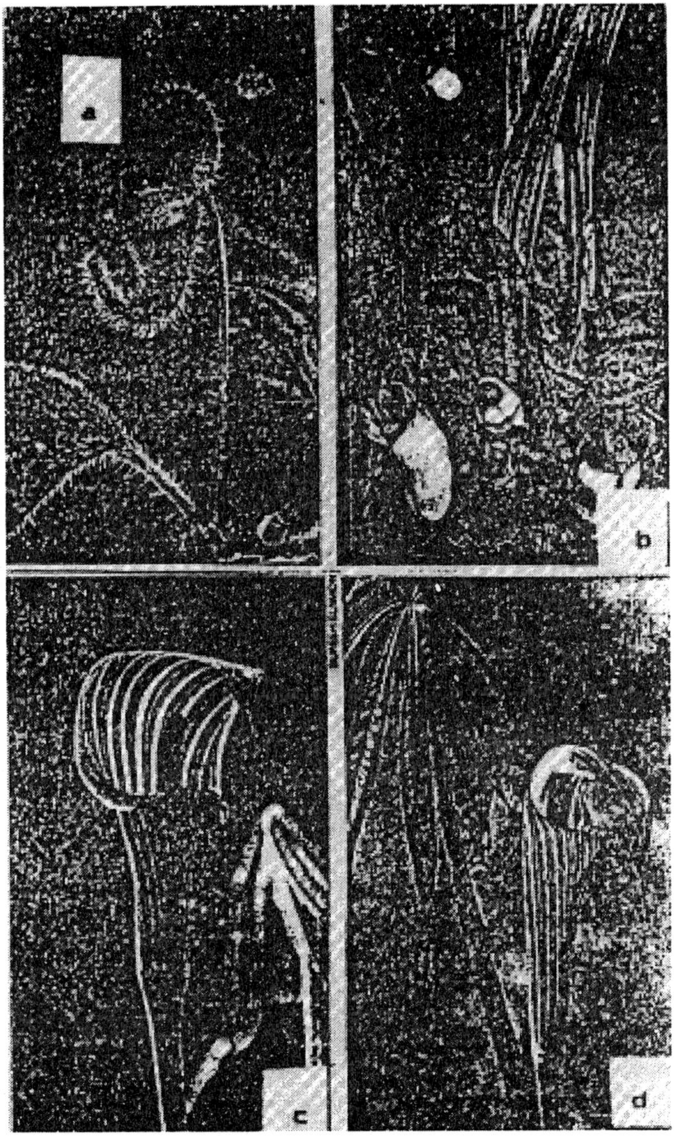

Fig. 32.2.a. *Impatients munronii;* b. *Acanthephiplum bicolor;* c. *Arlsaema leschenuaultii* (inflorescene with white and green spathe); d. *Arisaema leschenuaultii* (inflorescene with white and purple spathe)

are new to science. The TBGRI is currently using tissue culture techniques to multiply some hybrid orchid varieties.

Fig. 32.3.a. ***Dendroblum aqueum;*** **b.** ***Arsaema torusum*** **c.** ***Amorphophallus bulbifer***

The 21 plants unique to the Silent Valley are described and illustrated by S.D. Biju (who is the production editor of the book). His search took him back in time over a century, to accounts by Robert Wight, surgeon and botanist, who was the first foreigner to traverse the valley. Between 1840 and 1853, Wight published six volumes of his records entitled Icones Plantarum Indias Orientalis, documented in over 2,000 plates drawn by two Indian artists. Others who recorded the unique flora included W.G. Me Ivor, who created the Ooty Botanical Survey of India led efforts to collate information on the valley. Several people, including Professors B.K. Nayar and K.S.

Manilal of Calicut University, helped to publish detailed documents on the flora. Biju's paper provides an interesting account of his discovery. His search took him to the Royal Botanic Gardens at Kew in London. There he found Wights original sample of the rare *Impatiens iinmronii*. The plant was not seen thereafter, not until Biju located.

Fig. 32.4. Lion-tailed Macaque (*Macaca silenus*)

Another investigation fact revealed in the book is the existence in the valley of wild varieties of nearly 120 crops. Sheltered from pesticides, fertilizers and other artificial substances, these varieties have evolved new and robust strains, say K.S. Velaydhan and his colleagues from the National Bureau of Plant Genetic Resources. They have made a special study of the valley wild relatives of pepper and plead for the establishment of a gene sanctuary for these wild strains similar to the Garo hills of northern India, where a gene sanctuary for citrus was declared. Regular surveys by the

Forest Department and some students sponsored by funding agencies have established that the valley's animal population has thrived in the years since it was notified as a National Park. The Valley's most famous resident, the lion-tailed macaque (Fig. 32.4) according to KFRI estimates, now have around 275 members spread out in 14 troops (1996, figures) and an abundance of their favourite food, the fruit of the *Gullenia exarillata* tree has ensured that they do not migrate.

Large mammals—the elephant, the tiger, the sloth bear and the wild boar are also reported on the fringes and in the adjacent forests. The dense vegetation doesn't support many of these animals in the core of the valley, where lesser carnivores such as the civet, the jungle cat and the mongoose thrive. When it comes to birds, the book dramatizes the wonder that is the valley, by allowing P.K. Uthaman, field publicity officer by the job title and ornithologist by inclination, to recall his special connection with the valley. With the aid of set of stunning photographs Uthaman describes his quest to capture the nesting of the rare Malayan bittern, the shy peninsular bay owl resting among reeds, the crested serpent eagle on a deadly diving mission. The Valley harbours have 211 bird species. George Mathew's study of the butterflies and moths, admits that his records detailed as they are only scratch the surface.

b. Lion Tailed Macaque (Macaca Silenus) A Most Endangered Monkey of India

Introduction

The Lion-tailed macaque is an endangered primate species, endemic to the tropical rain forest of the Western Ghats. The total wild population is estimated less than 4,000 individuals distributed in the three southern Indian states of Kerala, Karnataka and Tamil Nadu. However, Silent Valley and Ashambu hills are the only two viable habitats in their entire range of distribution (Green and Minkowski, 1977).

Various stochastic events and catastrophes along with human interferences have been affecting seriously the Lion-tailed macaque population through the ages. Factors, such as habitat fragmentation, reduced habitat size, isolation of populations leading to inbreeding depression and vulnerability to random events make the Lion-tailed Macaque a highly endangered species despite its relatively large scattered population (Kumar *et al.*, 1995). Even if small fragmented populations facing severe problems of local extinction their counterparts existing in the large continuous protected areas like Silent Valley show better signs of population status (Ramachandran and Joseph, 1997). Population and habitat viability analysis workshop for the species held at Madras in 1993, strongly suggested the need for long-term monitoring of the species in its viable habitat. Such population monitoring data will definitely be helpful in the effective management of the species.

Silent Valley National park and adjacent areas are situated in Palghat District of Kerala state and located at 10°15' to 11°25' N latitude and 76°21' to 76°33' longitude. Silent Valley National Park is one of the prime habitats of the endangered Lion-tailed Macaque and Nilgiri Langur, both endemic to the Western Ghats of India. Silent Valley represent one of the core areas of Nilgiri Biosphere Reserve. The habitat is a tropical wet evergreen forest. Aiyar (1932) identified seven major tree associations in the area.

Status and ecology of Lion-tailed macaque was studied in Silent Valley National Park for a period of three years from 1993 to 1996 by Joseph and Ramachandran from KFRI, Peechi. This chapter is mainly based on their work published in Indian Forester, October 1998. Repeated census walks were conducted on foot through the existing trails and elephant paths randomly in the evergreen forests. Frequent steps were made to get the characteristic contact call of the Lion-tailed macaque in the vicinity of the transact. The located troops were followed and continuous observations were made until each troop was cited twice in a year.

Animals were classified as four major categories such as adult males, adult females, sub-adult males and immatures.

Adult males were identified with stouter body and long canines. The tail tufts were having slightly lifted carriage. Adult females were identified by the presence of elongated nipples and baggy breasts. Although, sub-adult males were having fully developed musculature and comparatively shorter canines. Rest of the individuals including sub-adult females, juveniles and infants were grouped into a single category as immatures. Demographic data were collected from 9 troops in 1993, 11 troops in 1994, 12 troops in 1995 and 14 troops in 1996. Of these seven troops were monitored continuously throughout the study period.

Results

Figure 32.5 shows location of various lion-tailed macaque troops in Silent Valley and adjacent areas. Fourteen distinct troops were identified from silent valley and adjacent areas when the field study ended in 1996. Among the fourteen different troops, eight of them ranged east of Kunthipuzha river and the rest towards the western side

Two troops each in Punnamala (PUN I; PUN II), Aruvampara (AV I; AV II), and Panthanthod (PAN I; PAN II) regions and one each in Sairandri (SAI) and Parathod regions (PAR) formed the resident troops East of Kunthipuzha river. Three troops in Chembotty areas (CHE IA; CHE IB; CHI-II) two troops in Nilikkal areas (NKL I; NKL II) and one troops in Puchapara (PUC) constituted the population west of river Kunthipuzha. All the troops were distributed at an elevation ranging between 700 to 1500 m.

Table 32.0 shows the status and population structure of Lion-tailed macaque troops in Silent Valley forests. The troops size varied from 9 (PAR) to 36 (NKL I) in 1996. The average troops size estimated was 19.64 individuals. Among the fourteen troops monitored in 1996, eight troops were having two adult males. But six troops had only one adult male. The number of sub-adult males varied from 1 to 3 in various troops. However, most of the troops had only one sub-adult male. Three troops were having no sub-adult males. The number of sub-adult males varied from 4 to 17 in various troops. Majority of the troops (9) had less than 10 adult females. Immatures ranged from 4 to 14 different troops with an average of 8 individuals. A total of 275 individuals were observed in the final census. Among these, 8 per cent were adult males and 45 per cent were adult females. The sub-adult males constituted 6 per cent and 41 per cent of the total population formed immatures. The adult sex ratio estimated was 5.64 females per male.

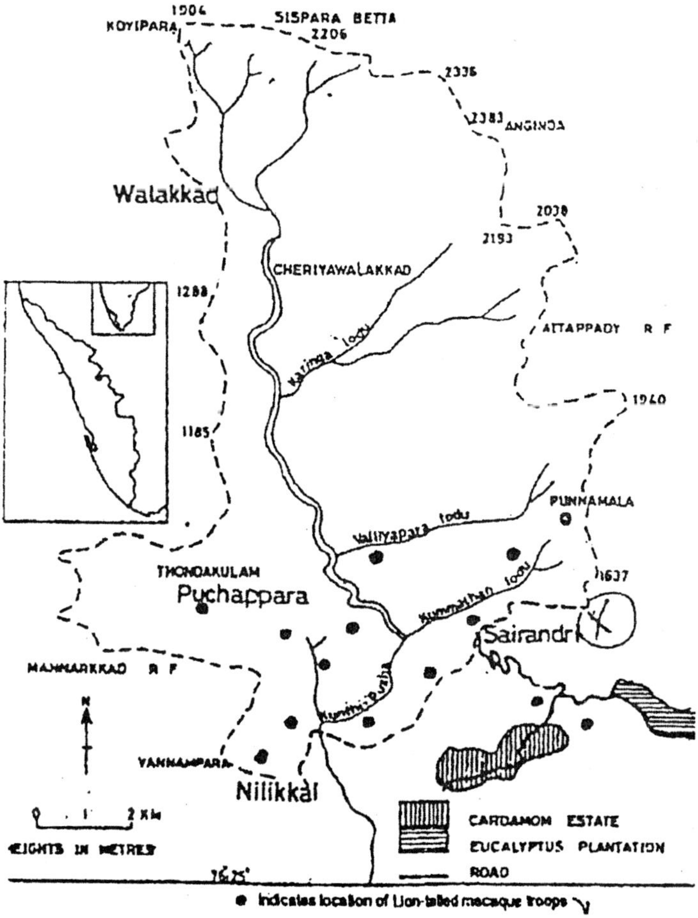

Fig. 32.5. Distribution of Lion-tailed macaque troops in the Silent Valley and adjacent areas.

Table 32.0. Population Density of Lion-tailed Macaque.

Troop name (Location)	*AM*	*SAM*	*AF*	*IMM*	*Total*
SAI (Sairandri)	2	3	16	13	34
PUN I (Punnamala)	2	2	10	6	20
PUN II (Punnamala top)	1	1	7	12	21
PAN I (Panthanthod)	2	1	9	8	20
PAN II (Panthanthod)	1	1	6	8	16
AV I (Aruvampara)	1	2	7	4	14
AV II (Aruvampara)	1		5	7	13
PAR (Parathod)	1		4	4	9
PUC (Puchapara)	1	1	6	8	16
CHE IA (Chembotty)	2	1	7	6	16
CHE IB (Chembotty)	2	1	10	7	20
CHE II (Chembotty)	2	1	11	7	21
NKL I (Nilikkal)	2	3	17	14	36
NKL II (Nilikkal)	2		9	8	19
Total	22	17	124	112	275

AM = Adult male, SAM=Sub-adult male, AF = Adult female, IMM = Immature.

Table 32.1 shows that the troop size varied across succeeding years in every troop. The mean troop size was greater in Nilikkal site, when compared to all other sites. The minimum troop size was found in parathod troop site. Majority of the troops size mean was 20 individuals. The overall animal number in all the troop was 19.24 animals.

Out of the nine troops monitored in 1993, five of them increased their total in 1994, while remaining three unchanged. There was a reduction in the troop size near Attapady from 21 to 13 in 1994, it is because the animals were killed by the Muduga tribals. A total of 11 troops monitored were increased their population size in 1995. Out of 12 troops monitored in 1996, two troops fissioned before 1996 census. Fifty per cent of the troops increased their number and remaining not changed.

Table 32.1. Year wise Variation in Population Density of the Lion-tailed Macaque

Troop name (Location)	*Troop Size* 1993	1994	1995	1996	*Mean*
SAI	29	29	31	34	30.75
PUN I	15	15	17	20	16.75
PUN II			21	21	21.00
PAN I	12	14	18	20	16.00
PAN II	21	13	15	16	16.25
AV	20	20	22		20.70
PAR	6	7	8	9	7.50
PUC	12	15	16	16	14.75
CHE I	29	31	33		31.00
CHE II		19	21	21	20.30
NKL I	28	30	32	36	31.50
NKL II		17	19	19	18.30
AV I				14	14.00
AV II				13	13.00
CHE IA				16	16.00
CHE IB				20	20.00
Total	172	210	253	275	19.24

Long-term demographic data on wild animals are lacking for many endangered and endemic species. This has handicapped the functions of wildlife management programmes. Lion-tailed macaque is being one of them and is most endangered species of India. It requires a special attention for its long-term survival. As most of the lion-tailed macaque habitats are fragmented, hence its conservation is most important. A number of surveys have been made to assess the population of lion-tailed monkey in different periods in the Western Ghats (Daniel and Kannan, 1967: Kurup, 1978; Green and Minkowski, 1977; Karanth, 1985; Esa *et al.*, 1997), very few of them were conducted for long-term. The first study on these animals was conducted by Kumar, 1987 in Varagaliyar forests. The present study is connected to the Western Ghats area of Kerala state, where more than 50 per cent of the population is located. In the silent valley areas monkeys are distributed between the altitudes 700 to 1500 m elevation. The occurrence of associated food plants viz., *Gullenia palaquium, Meusa palaquium, Meusa calophyllum,* may have influence on the distribution of lion-tailed monkeys.

The announcement of the area as a National Park in 1984 is a creative step and subsequent protective measures helped the survival of these monkeys. The low lying areas of Silent Valley have undergone drastic changes in the past and 1/5 of the areas became degraded grasslands due to forest fires. Trapping of nine individuals by the Muduga tribals in 1993 clearly indicates the vulnerability of these animals in the borders areas of the Park. Such stray incidents are restricted to the peripheral areas, not affecting the core forests of the park. The slow and gradual expansion of the macaque habitat during the recent years has positive implications for the population growth of the animals. This also envisages for the immediate steps for extending management strategies for long survival of the species.

Suggestions

1. The national park and adjacent areas are important habitats for the lion-tailed macaque. There is need to conserve the entire stretch of the adjoining rain forests for the long-term survival of these monkeys.
2. Fire control steps should be taken and monitored, continuously in the summer. Fire affected areas should be regenerated by planting the food plant seedlings for the monkeys and other primates.
3. The tourism zone be marked outside national park and anthopogenic pressure should be minimized in the core areas.

4. The monitoring of the endangered primates should be continued so that effective management actions will be taken regularly.

c. Legal Conservation of River Dolphins

The river dolphins are in great danger of extinction all over the world. The riverine habitat is highly degraded and is under heavy pressure everywhere as the human population is increased by leaps and bounds. Approximately 10 per cent of the human population of the world lives along the Yangtze river, the native domain of the Chinese river dolphin (*Lipotes vexilifer*) and industrial traffic on this river is geometrically increasing as China is building her future. Similarly, River Ganga is the home of Indian Gangetic dolphin (*Platinista gangetka*), is being greatly affected by the industrial and agricultural development in India, Bangladesh and Nepal. The Amazon and Orinoco River systems in South America are the habitats of American river dolphin are increasingly victim to the rapid loss of rain forests. If the present trend continues, there may be a little time to do anything to save and conserve several of river dolphins.

To cope up with these growing problems those countries have already passed the laws to protect river dolphins, but protection may not be possible because of the difficulty of enforcing these laws in remote areas and with limited resources. The present chapter describes the legal provisions to protect the river dolphins in various countries where river dolphins occur.

National Legislation concerning Protection of Habitat

Various types of legislations may be relevant to the protection of habitats of the dolphins. Among these are laws that establish protected areas such as parks and sanctuaries and laws that regulate the environmental implications of various developmental projects.

INDIA

The Wildlife Protection Act, 1972 was enacted for the protection of habitat, which also provides designation for sanctuaries. The Environmental Protection Act notification 1986 was also passed for the regulation and protection of environmental factors and handling the polluting and hazardous materials. This law also aims to protect our environment from the further degradations. There is also provisions to punish those who endanger the environmental safety and human well-being.

The new act (The Environmental Protection Bill 57-C/ 1986) appears to be re-written version with some modifications and deletions and is superior in its approach to the earlier ones. Unlike earlier acts Section 6(2) (b) of this act empowers the central government to make rules for the maximum allowable limits for various pollutants from different sources. Section 5 of the Act empowers the Central Government to issue directions for (a) the closure, prohibition or regulation of any industry, operation or process, (b) the stoppage of any other services such as water, electricity, etc. The penalties for defaulters in the new act are more stringent.

But this law does not specifically provide for a procedure to assess the environmental impacts of various industrial projects or other sorts of development. However, it is possible that this legislation could be amended to establish such a procedure. Environmental impact reviews could be useful for protecting river dolphins, as industrial pollution, water diversion and dam construction are considered to be major factors influencing the decline of Gangetic dolphins (*Platinista gangetica*) in India (Kumar *et al.*, 1995 and Kumar and Singh, 1996) since environmental pollution is considered to be a threat to *P. gangetica*, laws such as this may be relevant to river to dolphin conservation.

Pakistan

A river dolphin reserve was established in 1974 in the government of Sindh covering 170 km between the Guddu and Sukkur barrages on the Indus river. The sub-population in this stretch of the river is believed to contain over 80 per cent of all known Indus dolphins (Khan and Niazi, 1989).

The Pakistan Environmental Protection Ordinance, 1983 (Ordinance to provide for the control of pollution and preservation of living environment) requires that an environmental impact statement must be submitted to the Pakistan environmental protection agency by every proponent of the project, the construction and completion of which is likely to adversely affect the environment. The impact statement must include the adverse environmental effects and the steps proposed to be taken to minimize these effects. The agency may involve the public participation in the assessment of the environmental impact statement. The agency may recommend to the federal government that the project be modified or rejected in the interest of environmental objectives information on whether agency has issued regulations for carrying out the purposes of this ordinance was not available.

BANGLADESH

The Third Schedule of the Bangladesh Wildlife (Preservation) Order, 1973 provides for the establishment of sanctuaries but till date none has been designated.

BHUTAN

Although, no information on habitat protection laws was available, there exists a large sanctuary adjacent to the Indian border that might protect the river dolphin habitat.

CHINA

The People's Republic of China Environmental Protection Law was enacted to ensure rational use of the natural environment, to prevent the environmental pollution and ecological destruction and to promote economic development. The environment as defined in the law includes water and aquatic organisms. Chapter 1 defines the principles of the law. Whenever projects are to be undertaken, including new project or those involving the rebuilding or expansion of existing ones, an environmental impact report shall be submitted to the Environmental Protection Department and other relevant departments for examination and approval before any planning is commenced (para-6).

This law and other statutes and regulations relating to Environmental Protection apply to foreigners (Para-9).

Chapter-2 describes measures to be taken to protect the environment. Among those applicable to the river dolphins are: Comprehensive scientific surveys must be carried out before the new construction of small, medium or large-scale water works, etc. Measures for the protection and improvement of the environment shall be conscientiously adopted to prevent the destruction of the ecological system (Para-10).

Aquatic or land organisms must be protected, developed and rationally used. Over harvesting and destruction causing extinction must be prohibited (Para II).

However, it is unclear whether the river dolphins are considered to be aquatic organisms or not.

Chapters-4 and 5 describe the duties of governmental agencies, including international co-operation, environment monitoring, scientific research and environmental education.

ARGENTINA

Argentina passed a law in 1980 to protect certain area. In addition to National Parks, these include Natural Monuments, defined as regions, object or fauna and flora of aesthetic, historic or scientific interest should receive complete protection. National Reserves may be established to conserve ecological systems.

VENEZUELA

This country has created protected areas by Executive Decree, under the Forest, Soils and Water Law of 1965. The Organic Environmental Law of 1976 establishes the importance of conservation, to be achieved through the creation of sanctuaries, wildlife refuges, forest and hydrological reserves, national parks, monuments, etc. Over 7.6 million hectares area are protected and 280,000 hectares are totally protected as faunal refuges. If destruction of aquatic and forest habitat is a threat to river dolphins, existing legislation may provide some measures of protection. Sections of Venezuela's forestry and water laws pertaining to activities that would destroy the vegetation for agricultural purposes (vide Decree 1804 of 20th January, 1983, and Resolution No. 491 of 14th April, 1983).

Colombia

Colombian National Park Legislation contained in Law 2a of 1959, Decree 2420 of 1968, Decree Law 2811 of 1971, Decree of 16 March, 1977. Among the types of protected areas established under the present legislation are National Parks, Natural Reserves and National sanctuaries. Law 30 of 1980 approves the agreement for the conservation of flora and fauna of the Amazonic territories of Colombia and Peru, which was signed in Lima on 30 March, 1970.

Peru

Decree to Ley No. 21147, authorizes the establishment of national parks and other conservation areas. Total protection is afforded I *ma* and *Solatia* in the Pacaya National Park in the Upper Amazon basin.

Brazil

Lei No 4771, 1965 is Brazil's major forestry law, which applies to the preservation, exploitation and management of forests. This law provides scope for the establishment of National Parks, and Biological Reserves where modification of any type is prohibited except for authorised scientific purposes. Biological reserves are created for safeguarding, exceptional natural qualities, flora and fauna for educational and scientific purposes.

Lei No. 6938, 1981 which adopts provisions concerning national environmental policy, may provide a means to protect river dolphin habitat. Among the stated purposes are preserving and improving the environmental quality and assuring conditions to protect the dignity of human life. The law creates the Cadastro IVi nico Federal de Ativades Instrumentos de Defesa Ambiental for the mandatory registration for individuals and corporations devoted to technical advice on ecological and environmental problems and of industry and commerce in equipment and instruments intended to control polluting activity. Population from agricultural, mining and other industrial activities may become cause for concern for *Inia*.

NEPAL

According to Shrestha (1993), the river dolphins (*P. gangetica*) in Nepal have been caught traditionally for meat and oil, which is used for burning and medicinal purposes and to attract fish. Hunting of river dolphins is prohibited by the National Park and Wildlife Conservation Act and punishable by a fine and or imprisonment. The act also established the Royal Chitwan National Park, where *P. gangetica* is occasionally sighted. All wildlife exploitation is prohibited in the national park. The dolphin is also found in the Royal Karnali Wildlife Reserve. Klinowska (1980) reports that unauthorized fishing in the Chitwan Park and legal fishing in the Wildlife Reserves may pose problems inspire of legal protection afforded to the dolphins.

BOLIVIA

Bolivia's principal wildlife law is Ley de Vida Silvestre, No. 12301. An implementing decree, Decreto Supremo No. 16605 June, 1979 declares a total ban on the hunting and trade of listed native wildlife species considered to be in danger of extinction. Among the species listed is *Inia geoffrensis*. According to Fuller and Swift (1985), the validity of the decree has been questioned because of some scientific errors in the listings. *Inia geoffrensis* is erroneously classified under pisces.

ECUADOR

The 1981 Ley Foresataly de Conservation de Areas Naturales Vida Silestre (Ley No. 74) governs conservation and use of living resources except fish and marine species. There are some inconsistency between the law and its implementing decrees (No. 1529) regarding export provisions. According to the law, export is only permissible for scientific or educational purposes. In the decree, these restrictions only apply to protected species. It is unclear whether the river dolphins are specifically protected. Presently however,

Ecuador prohibits the commercial export of indigenous species.

GUYANA

Under Guyana Fisheries Regulation (Section 30) Act, the hunting of a few aquatic species is controlled. The regulations require that the hunting or capture of listed species be done under a license from the fishery department of the ministry of agriculture. *Inia* could presumably be listed on the schedule of regulated species by administrative action.

AMERICA

The U.N. National Marine Fisheries Resource is formally considering petitions to list *Lipotes vexilifier* as an endangered species under the U.S. Endangered Species Act (USESA). The provisions of the USESA apply both to trade of river dolphins and to protection of their habitat. The U.S. Marine Mammal Protection Act (USMMPA) also inhibits the import of marine mammal parts in the U.S.A. It also contains a directive to encourage the international agreement for the protection and conservation of marine mammals. Another U.S. law that is National Environmental Policy Act (NEPA) scrutinizes all the major developmental projects before implementing and it also give protection to aquatic mammals.

International Agreements and Programmes

During the past 3-4 decades the earth is facing environmental degradation problems in terms of increased greenhouse gases, global warming, depletion of tropical forests, pollution of air, water, soil and extinction of plants and animals. These have created unpleasant changes in climate, sea level, monsoon, winds and ozone layer. Some important programmes undertaken to protect the wildlife are mentioned herein.

Western Hemisphere Convention, Washington, 1940

Argentina, Brazil, Chile, Ecuador, Peru, Uruguay and Venezuela formed a party and arranged a convention. The main objectives of the convention is to preserve the native flora and fauna from extinction and to preserve the areas and formations of extraordinary aesthetic, historic and scientific values. The convention provides scope for the establishment of National Parks and Wildlife Sanctuaries and of strict wilderness areas. The parties agreed to adopt laws to protect and preserve such species and to regulate their trade. Species needed special protection are listed. The only Cetaceans listed are four species of great whales. This convention might become useful for the conservation of *Inia* also.

International Importance of Wetlands

This was conducted in Ramsar in 1971 to focus the recognition of the ecological as well as economic, scientific, cultural and recreational importance of wetlands. The Ramsar convention aims to curtail the loss of wetlands. The convention has proved to be useful tool for preventing deleterious changes occurring in wetlands. Although, the emphasis is on the protection of wetland birds and their habitats, many of these areas are also important for river dolphins particularly *Inia*.

World Heritage Convention

This convention was organised by the countries namely Argentina, Bangladesh, Bolivia, Brazil, Colombia, Ecuador, Guyana, India, Nepal and Pakistan in Paris in 1978. This convention was concerned to the protection of world cultural and natural heritage. The main purpose of this society was to protect precisely delineated natural areas of outstanding universal values from the point of view of science conservation. The parties agreed to develop scientific studies and take necessary legal scientific steps to protect their natural and cultural heritage.

Convention on International Trade in Endangered Species

This convention was held in Washington in 1973 to prevent the over exploitation of flora and fauna due to international trade. This summit developed a system of trade control by passing many resolutions. Species needed special protection are listed in the CITES appendices. Species threatened with extinction are listed in appendix 1. and the *Lipotes vextlifer* and *Platinista* are included in this list.

Convention of Flora and Fauna of Amazon Territory

This agreement took place between the two countries, Peru and Brazil in 1975. It aims to preserve flora and fauna of Amazon territories.

Bonn Convention

This was held in Germany for the conservation of migratory species of wild animals. The objective was to protect the species that migrate across and outside national boundaries. This treaty is potentially useful for addressing river dolphins.

Brasilia Treaty

This summit was attended by the representative from Bolivia, Brazil, Colombia, Ecuador, Peru, Venezuela, Guyana and Surinam in 1978 for Amezonian cooperation. Although the treaty is concerned primarily with socio-economic concern, it also recognizes the need for rational use of Amezonian flora and fauna with an objective of maintaining the ecological equilibrium of the region and preserving the species. The treaty promotes scientific studies among the parties.

Lima Agreement

This meeting was organised by Peru and Colombia in 1979 to conserve the diversity of Amezonian region. Under this agreement certain organizations were assigned the function of promoting and putting into practice the scientific interchange and other measures to ensure proper management of natural resources and protected areas.

Man and Biosphere Programme

Man and Biosphere (MAB) aims to establish protected area network for endangered species. MAB uses existing national and international legal and administrative mechanism. MAB is admitted by the UNESCO. Several of the countries where dolphins still occur have already established biosphere reserves and those countries are Bolivia, Colombia, and Peru.

SAARC Summit

South Asian Association for Regional Cooperation (SAARC) is an organisation of seven south Asian countries viz., India, Pakistan, Nepal, Bangladesh, Bhutan, Sri Lanka and Maldives Islands. Its objective is to strengthen regional cooperation among the partner countries for the welfare of their people and to maintain eco-balance of the environment.

Earth Summit

Though the first earth summit was arranged in Stalkholm in 1972, but the officially announced and well-known U.N. conference on Environment and Development (UNCED) was held at Rio de Jenerio (Brazil) from June 3 to 14, 1992. Heads and representative from 178 countries participated in the discussions to adopt global plan to save the earth from degradation. This summit also resolved the protection of threatened and endangered flora and fauna of the world. The main aim of the summit was to act quickly and wisely on the problems of eco-degradation of the biosphere.

Status of Indian Dolphins

The Gangetic dolphins are under potential threat of extinction because of industrial pollution and other stress from construction of dams, diversion of river flow, navigation, fishing and poaching activities. If the present trend of eco-degradation of the Ganga continues, this species will gradually disappear and may become extinct. The government of India has also planned to fund the population research on the dolphins in the Ganga waters. A number of workers (Peterle, 1932; Mohan, 1989; Jones, 1992; Kumar, 1995) have studied the problem and made various suggestions.

Recommendations

Establishment of statutory bodies based on the following laws and agencies is required :

1. International Environmental Protection Authority.
2. Environmental Impact Assessment and EMPS groups.
3. Hazard and health damage compensation councils.
4. Wild life Protection Act.
5. Various NGOs involved in the wild life protection activities.

These agencies will serve as watch dogs for the protection of environment and also assist national governments and courts regarding the environmental issues.

d. Gir Forest: A Last Home for Asiatic Lion (Panthera leo persica)

Gir is a single largest tract of forest in the Saurashtra region of Gujarat state and is synonymous with the Asiatic lion, *Panthera leo persica*. The lion might have entered India from the west and was found in large numbers in the states of Punjab, Haryana, Rajasthan, Uttar Pradesh, Madhya Pradesh, Gujarat and Jarkhand states. The last lion surviving outside the Saurashtra was killed in 1884. The Gir forest in Gujarat has shrunk from 3070 sq. km to 1884 sq. km in 1880, due to expansion of agriculture and destruction of habitat. In the early part of this century the Gir forest was connected to Girnar, Mitiyala, Barda, Alech hills, Dhank and Chorwad by corridors of rough semi-wooded forests, grasslands and sparsely populated villages. This made the lions to move freely in the area.

Lions disappeared from Barda and Alech hills towards the mid of 19th century and disappeared from Girnar and Mitiyala by 1963 and 1965 respectively. At the time of notification of Gir forests as a lion sanctuary in 1965, lions were found only in the compact forest of the Gir. Gradually, the size of wildlife protected area was increased to 1412 sq. km with the national park surrounded by the sanctuary. The rest of the 403 sq. km peripheral forests constitute the buffer zone of the wildlife reserve. The most important aspect of the Gir forest is that it has become a very stable ecosystem with tremendous regenerating, self supporting and sustaining capacity due to its rich and diverse flora and fauna. The Gir is an unique ecosystem with over 4000 species of plants, 32 mammals, 24 reptiles, 310 birds and more than 2000 insect species and other invertebrates.

Conservation Through Single Species Project

A study by Paul Joslin (1972) revealed that Gir displayed a sadistic evidence of accelerated degradation of the ecosystem, and he has warned that if nothing was done to arrest the rate of decline in the number of surviving Asiatic lions, the species would become extinct within two decades. Recognizing the serious danger to the lion and pitiable conditions of Maldharis, the Gir lion sanctuary project for initially five years was prepared in 1972. It was implemented at a total cost of 58 lakhs. The work carried out under the Project Lion on the following lines:

1. Fencing of the sanctuary by constructing a rubble a wall of 400 km along its periphery.
2. Constructing barricades along the water course traversing the periphery of the sanctuary.
3. Establishing checkposts across all public roads, passing through the sanctuary.
4. Shifting of all maldharis families from the national park and majority of the families have been shifted outside the Park, who are given cultivable lands, grazing lands, and housing sites. Out of the 845 maldharis families, about 2/3 have been moved and re-settled outside the park at different places.

Serious problems like grazing, poaching, lopping and cutting of trees, etc., were tackled by the Department of Forests and development programmes like, soil and moisture conservation, and plantation, etc., were implemented. The habitat conditions improved and in many places of the Gir become thick after the launching of the project. Constructing of water harvesting structures and artificial water holes has increased the availability of water to the wildlife.

Positive Changes in Wildlife Density

The implementation of Gir lion sanctuary and park has improved and increased the availability of food and water. Six consecutive census have been carried out at an interval of five years after launching the Project Lion. The lion population along with other wildlife has increased consistently (Table 32.2).

Table 32.2. Growth of Wildlife Numbers in the Gir Park

Year	*Lion*	*Leopard*	*Hyena*	*Ungulates (Herbivores)*	*No. of Ungulates lion*
1968	177	NA	NA	NA	NA
1974	180	142	63	9635	53.5
1979	205	161	84	14964	73.0
1985	239	201	192	16905	70.7
1990	284	212	97	31489	110.9
1995	304	268	137	38221	125.9

NA = Not available.

The herbivores include six major species namely Chital, Sambar, Blue bull, Chinkara, four horned antelope and wild boar. There has been a consistent increase in the carnivores and herbivores. Availability of ungulates per lion has increased from 53.6 in 1974 to 125.7 in 1995 which makes wild kills, easily available to carnivores. The study of Kavi Challam indicated that the food pattern of the lion has changed in favour of wild prey as 65 per cent of the lion diet was recorded as wild animals where as 35 per cent was made up of livestock kills.

Group Structure

Lions are social animals usually live in families and hunt in groups. The size of the group recorded in the Gir was around a dozen. 304 lions were located at 94 sites on live bait in May 1995, where as the number was 284 in the previous census of 1990.

Migration of Lions

Lions disappeared during middle of this century from the forests outside the Gir. The habitat of the Gir improved and the lion population gradually grew with the increased level of protection and conservation. Before the census in 1990 lions were occasional visitors to Girnar, Mitiyala and coastal forests, but they started visiting the neighbouring forests again. The situation has changed then and this carnivore starts moving into the former territories. At present, there are four satellite populations of lions. The second generation of migrated lions have made Girnar and coastal forests as their home range. The number of lions outside the Gir are treated as floating population because some of them visit the Gir zones for short periods and go back to their home range. In all the four zones the herbivore count is low therefore the migrant lions primarily depend on domestic livestock kills. It has been observed that the majority of the migrated lions were sub-adults probably they were compelled by the dominant males to leave the pride and the territory. A long strip of coastal plantation supports a small population of blue bull and wild boar. One pride of 16 lions in Kodinar Dhamrej Sutrapada and another 10 lions in Rajula and Junabad depends on cattle and blue bull. Lions were brought back to the Gir in the past but they repeatedly migrated to coastal forests in search of food and territory and settled in those areas permanently after 1990. The fourth satellite population of lion is in the Mitiyala forests of Bhavanagar forest division, and is in the process of settling. The Gir lion population is showed in Table 32.3.

Table 32.3. Distribution of lions in different areas

Sl. No.	*Site*	*Male*	*Female*	*Cub*	*Total*
1.	Gir	92	109	61	262
1.1	National Park	10	12	9	31
1.2	Sanctuary	72	92	50	214
1.3	Peripheral area	10	05	02	17
2.	Girnar	10	03		13
3.	Mitiyala	02	01		3
4.	Kodinar Sutrapada coastal forests	05	06	05	16
5.	Rajula-Junabad area coastal forests	03	02	05	10
		112	121	71	304

Carrying Capacity

The carrying capacity of an area depends upon the availability of food, water and conduciveness to reproduction and space. Food is not a major limiting factor in the Gir as the population of important ungulates increased at the rate of 14.2 per cent per year during the last three decades, but shortage of space may resulted in territorial fighting and migration. The population has remained almost the same level, it was 2d7 in 1990 and 2h2 in 1 Wi in the National Park and Sanctuary during the past HI years.

The natural dispersion of lions started since the last drought in 1987. 17 lions were recorded outside the Gir in 1990 which increased to 42 in 1995 but the population remained constant in the PAN. The present number of lions in the PAN indicates its carrying capacity. The prey base available during 1995 in the Gir protected area shown in Table 32.4.

Table 32.4. Prey base available in the year 1995 census in Gir protected areas

Species	*Common name*	*Population*
Panthers leo persica	Lion	304
Panthera pardus	Leopard	268
Hyaena hyaena	Striped hyaena	137
Axis axis	Chital	32061
Cervus unicolor	Samber	2262
Boselaphus tragocavme/us	Blue bull	1856
Gazelle gazelle	Chinkara	441
Tetraceris quaadsicornis	Four horned antelope	387
Sus scrofa	Indian wild boar	1214

Conservation Strategy

Greater Gir ecosystem or Extended Gir in the new territories of lions has become the need of the hour in the interest of the increasing population of lions. Proper strategies to expand the Gir present area of 1814 sq. km to 2370 sq. km.

in the near future by covering Girnar (179.5 sq. km), Mitiyala (19.4 sq. km), Coastal forests (110.1 sq. km), Barda (187.4 sq. km) and Vidis (60 sq. km) areas. In addition to the forest lands, wastelands and panchayat lands connecting Gir, the above forests should be developed and managed as corridors for lions. The following measures are required for the management of the habitat:

1. The habitat of the Gir should be improved and expanded by arresting the degradation of peripheral forests and wastelands.
2. The Maldharis still living inside the sanctuary and park should be attracted to re-settle outside.
3. The satellite population at the four sites should be managed by increasing the prey base.
4. The Girnar and coastal forest should be developed and the herbivore population should be managed in those areas.
5. Barda forest which was considered as the alternative home for the lions should be rehabilitated and prey base for the lions should be increased.
6. For herbivores, in-situ breeding programme should be developed to increase the ungulate population in Girnar, Mitiyala and the coastal forests.
7. Necessary management practices should be followed to facilitate natural migration of lions in new areas. The problematic animals should be utilised for the breeding purposes in captivity.
8. Management of Girnar, Barda, Mitiyala, coastal forests and other neighbouring forests with the management of the Gir should be integrated. The concept of greater Gir ecosystem management should be evolved, not only for the conservation of asiatic lion, but also to provide ecological security to the region. Management of wasteland, panchayat land, grassland and forests connecting Gir to new

habitats should be covered under a management plan as Greater Gir ecosystem.

(e) Musk Deer-(Mosclius chrysogaster)-Musk Live Extraction

The Himalayan musk deer-*Mosclius chrysogaster* is primitive ruminant commonly found in the Himalayan region between 2400 to 4300 m in Nepal. The main threat to musk deer is its musk extraction. The musk sac is located between the reproductive organs and the umblicus. It opens to the exterior through an orifice which lies anterior to that of urethra. Traditionally the musk pod is incised out after killing the deer. The musk can also be extracted by simple operation. But this method is time consuming and causes much stress in the deer.

Mukti N. Shrestha (1988) has explained a simple method of musk extraction from live deer and it is as follows. An 18 months old male was physically restrained and secured to expose the umblical region. A silver scoop with one large and one small groove at each end was sterilised and umbricated with antibiotic cream. Holding the musk sac with the left hand, the scoop was inserted gently with a rotating movement. The edge of the scoop is smooth and rounded to facilitate easy insertion and to prevent injury to the musk gland. This method was repeated at different ages of the buck for the musk and it was extracted and analysed for its quality (Table 32.5).

The length of the musk sac was 4.1 cm externally and width and depth were 3.3 and 4.6 cm respectively. At the age of 1.5 years the extracted musk was not of the desired quality with a pasty appearance and pungent urine smell. Musk at the age of 2.5 years was of good quality. The brown granular powder was strongly scented. At 3.6 years the musk extracted after 26 days after previous extraction was comparatively less in yield (Table 32.5). Musk yield varies at different age because of the health conditions and frequent extractions in small quantities previously. The musk extracted during 3rd week of December and January was better in terms of the quality, whereas the musk extracted during November, April and June was immature. The musk can be extracted annually during

Table 32.5. Extraction of Musk and its Characteristics

	Date of musk extraction and age of Musk Deer	*Musk yield in grams* *Wet*	*Dry*	*Moisture content in %*	*Colour, consistency, scent*	*Remarks*
1.	20 Dec. 1981 1 year 6 months	2.000	0.325	83.75	Dark red-brown, pasty, urine smell	Extracted at beginning of expected mating season (December-January).
2.	23 Dec. 1982 2 years 6 months	5.700	5.655	0.78	Dark red-brown, granular strongly scented	-do
3.	19 Jan. 1984 3 years 7 months	2.200	1.150	2.27	-do	Extracted 26 days after previous extraction.
4.	6 Dec. 1984 4 years 5 months	–	–	–	Light brown, pasty slightly scented	Small amount extracted to check colour and consistency 3 weeks before first extraction
5.	11 April 1985 4 years 9 months	–	–	–	Creamy white, pasty no remarkable scent	Small amount extracted to check colour and consistency 5 months after expected mating season.
6.	21 Dec. 1985 5 years 6 months	3.000	2.380	20.66	Dark red-brown, granular, strongly scented	Extracted in expected mating season.
7.	27 June 1987 7 years				Creamy white, pasty, no remarkable scent	Small amount extracted to check colour consistency and scent 6 months after mating season.

the rutting season by this method from adult males in the wild after catching thorn with nets and they may be released after the harvest of musk. This particular method will avoid killing the buck, which create imbalance in the sex ratio and endangering the animal to extinction. In this method, the local village folk may be involved and they may be given the benefits, so that the villagers themselves will become watchdogs against poaching and ensuring the survival of the endangered musk deer even in India and elsewhere.

f. Project on Great Indian Bustard Ardiotis nigriceps

Great Indian bustard is one of the rarest birds of the world. It was existing in the short grass plains of Uttar Pradesh, Tamil Nadu, Sindh area of Pakistan, and in the west of Orissa. Unabated shooting and destruction of its habitats exterminated the bustard from most of its range. By 1980, it survived only in six states of India in small populations. These states include Madhya Pradesh, Rajasthan, Gujarat, Maharashtra, Andhra Pradesh and Karnataka. The Tourism and Wildlife Society of India has organized an international conference on Bustards in 1980 at Jaipur, which initiated special conservation project on Great Indian bustard. It has also alerted all the states to undertake conservation measures and eight protected areas for the bustard have been established (Rahamani, 1987, 1989).

In the year 1980, the bustard population was estimated to be around 500-1000 with half of them surviving in the Rajasthan (Rahamani, 1993). Two extensive surveys made by Frot Rahamani and his team in the Thar desert and regular visits to bustard sanctuaries in Madhya Pradesh, Gujarat, Andhra Pradesh showed that bustard number is drastically reduced in Madhya Pradesh, Gujarat and Rajasthan while the number was stable in Rollapadu of Andhra Pradesh and increased in Nannaj (Maharashtra-Table 32.6). Besides the Great Indian bustard, three more species of bustards also occur in India. They are the Houbara (*iTi/d/m/r/o/rs iintffilata*), the Lesser Florican (*Sypheotides indica*) and the Bengal Horican (*Houbaropsis bengalensis*).

Table 32.6. Population Estimates of Bustard in the 1980s and 1992-93

State	*Area*	*Earlier numbers*	*in 1992-93*
Madhya Pradesh	Karera	20-25 (1983-86)	4-5 (1993)
	Ghatigaon	12-15 (1983-85)	3-5 (1993)
	Phori	10-15 (1983-84)	Not surveyed but few still survive
	Panna	5-10 (in 1988)	Not surveyed
Gujarat	Bhatiya	5 seen (1984)	None seen in 1993
	Lala	1 seen (1984)	Sin 1981
Maharashtra	Nannaj	10-13 (1981)	40.45 in 1991
	Karmala	8 (1981-82)	None
	Kamuni	5-8 (1981-84)	3 in 1992
Andhra Pradesh	Rollapadu	50(1987-88)	50(1992-93)
Karnataka	Rannibennur	5-10(1984-85)	5-6(1991)
Rajasthan	Sorsan	8-10(1984-86)	8-10(1991-92)
	Sonkhallya	30 seen in 1986	None in 1993*
	Diyatra	13 in 1986	4 in Feb. 1993
			None in July 1993
	Bap	11 in 1983	6 in Feb. 1993 None in July 1993
	Sam-sudasari	25-30 in 1986	5 in Feb. 1993
			15 in July 1993
	Khuri	14 in 1986	None in 1993
	Miyajlar	Sin 1986**	None

Source: Rahaman 1993

The Project area includes Thar desert, Kutch, Solapur Ahmedanagar districts of Maharashtra, Kutch area of Gujarat, Rollapadu (AP), Karera and Ghatigaon reserve areas (MP). The survey revealed the following status of bustards in different places.

Madhya Pradesh

The bustard population declined all over its range in Madhya Pradesh. The greatest decrease was seen in Karera and Ghatigaon bustard sanctuaries which were specially established for the protection of this species. In Madhya Pradesh, there were four known bustard areas; Karera, Ghatigaon, Pohri and Parma. In the mid 1980s, the total bustard population in these four areas was estimated to be around 50, with Karera and Ghatigoan having 30-35 birds. In the late 1980s, there has been a progressive decrease of bustards in these sanctuaries. In Karera, not more than five-birds were found. Even territorial males which were seen for many years and as villagers remember, were not seen during the display season in 1993. Condition of Ghatigaon was as bad as Karera. The Forest Range officer in-charge of bustard at Ghatigaon saw only two birds in six months.

Gujarat

In Gujarat, by 1980s the bustards had become extremely rare and survived only in two districts i.e., Bhatiya in Kalyanpur taluka of Jamnagar, and Abdasa and Mandvi talukas of Kutch. There was stray record from Surendranagar district. Total population in the whole of Gujarat was estimated to be between 20 and 30 bustards. During the last ten years, bustard, number declined by almost fifty per cent. In 1990, a small bustards sanctuary of 100 ha was established near Lala village where breeding was noted. In the Bhatiya area of Jamnagar district, where the Gujarat Government is planning a bustard sanctuary five birds were seen in 1983. It is doubtful if the population can recover from such low numbers.

Maharashtra

Nannaj area in Solapur district of Maharashtra has now the most satisfactory increase in the bustard numbers, mainly because the Forest Department provided good measures to protect its grassland habitat. In 1981, the maximum number seen was only eight. Since then successful breeding has been observed and by August 1993, there were 37 bustards. This increase was due to successful breeding and immigration from surrounding areas.

Andhra Pradesh

Rollapadu bustard sanctuary in Kurnool district has also shown positive result, thanks to effective protection to the bustard and its grassland habitat, especially during the initial stages of establishment of the sanctuary. Territorial males and successful breeding were sighted every year. However, there appears to be some laxity in protection when the birds go out, hence the increase in the bustard number is not true as it is projected. Reports of bustard shooting by rich vehicle borne poachers from Hyderabad could be one of the reasons why the bustard numbers have not further increased. During 1987-88 around 50 bustards were estimated in the Rollapadu grasslands, and the present estimate (in 1993) is also the same.

Karnataka

In Karnataka bustards are chiefly seen around Ranebennur, Blackbuck Sanctuary and Guttal plantation in Dharwad district. However, as the sanctuary itself is not very suitable for bustards due to excessive growth of eucalyptus, the birds are seen outside the sanctuary in grazing land which suffer from over-exploitation (Fig. 32.6).

Rajasthan

During 1980s, more than half of the bustards in India were present in Rajasthan, mainly in the Thar desert (Rahamani,·1989; Rahamani and Manakadan, 1990). In nine districts; Kota, Ajmer, Bhilwara, Jalore, Pali, Bikaner, Jodhapur, Jaisalmer and Barmer (Vardhan and Goriup, 1980). Now, all over the Thar desert, bustard population has drastically declined (Table 32.6), e.g., Diyatra, Bap, Sam, Sudasari, the bustard numbers have halved or disappeared as in Khuri, Miyajlar.

The bustard is still reported from a large area in the western Thar desert (mainly Jaisalmer district) but in Nachna, Ramgarh, villagers reported bustard poaching by outsiders.

Most of these rich poachers hunt for Houbara and Sandgrouse, but also shoot the Great Indian Bustard.

The most surprising decline was seen in Sam-Sudasari areas of the desert National park in Jaisalmer under Forest Department, with better protection than other core areas (e.g. Miyajlar, Khuri, Sotto) of the Park, where the population between in 1986 and 1993, reduced from 25 bustards to five. Forest Department guards however, claim up to 125 bustards.

Fig. 32.6. Great Indian Bustard (*Ardiotis nlgrlceps*)

Reasons for decline

1. Destruction of breeding areas

Destruction or alteration of grassland habitat is perhaps the most important reason of the decline of bustards in Madhya Pradesh and Gujarat. Conservation success depends on protection to adult birds and to their breeding areas (Table 32.7).

2. Disturbance

There is no bustards area in India, which is free from human disturbances. Like grazing, cultivation and their activities linked to increase in human population are listed in Table 32.7.

3. Shooting

At Karera, Ghatigoan, Nannaj, etc., poaching of bustards is under control but shooting appears to be playing a major role in decimation of bustards of the Thar desert. All the four species of bustards found in the India have the highest legal protection but implementing the law more stringently to eliminate poaching is crucial. The habitats of the bustards are not well protected and this constitute the biggest threat to bustards. Therefore, our major conservation efforts have to be addressed to habitat protection especially breeding grounds.

Table 32.7. Specific Disturbances to Bustard Sancturies/Areas

Name	*Type of disturbance*	*Migratory measures*
Karera	Construction of canal	None
Ghatigaon	Construction of railway line and canal	None
Pohri	Construction of gas pipelines	Compensation
Nannaj	Plan of a Spinning Mill	Plan cancelled
DNP	Construction of a canal	None
Lala	Prosopis plantation	Plantation stopped

Source: Rahaman 1993.

Project Bustards

The grassland and desert are under represented in the protected area network in our country. The studies by Rodgers and Panwar (1988), on endangered species Project and grassland ecology project of the Bombay Natural History Society have identified many potential areas which could be developed as new sanctuaries. There is a need to have a nationally co-ordinated conservation project for the Indian bustards and their habitats. Hence a project bustard is suggested.

g. Nesting Ecology of Baya Birds in the Western Ghat Regions of Karnataka

Introduction

Birds can live in different habitat conditions and these have been blessed with the plumage and flying capacity. The specific aerial adaptation made these birds as the most successful species on the earth. Perhaps, due to this reason even under habitat loss the birds are comparatively less affected when compared to other species of wildlife.

In the evolution tree of the animal kingdom, next to insects, birds are the important animals adopted to live in all types of habitat. These are the homeothermic animals being spread in different parts of the world in different habitat conditions. There are more than 8650 genera of birds have been identified and studied on the earth. In India, 1200 species of birds reported from various parts, which include 300 species of migratory birds visiting every year during winter. Late Dr. Salim Ali has classified Indian avifauna into 71 families.

The general conservation and management of birds encompasses the aspects like how do men trouble the birds, how do birds harm men and what is the possible measure to solve the problems. Though, India has not lost any bird species, but according to IUCN red data book 6 birds are endangered, 6 are vulnerable, 5 are intermediate, 4 are rare and one is extinct from Karnataka respectively (Hosetti and Venkateswaralu, 2001).

According to *The Book of Indian Birds* (1996), the common baya or Indian baya (*Ploceus philippinus*) is widely distributed all over India, however its ecology has been studied only in Gujarat, Rajasthan and Andhra Pradesh. Salim Ali reports that *Ploceus bengalensis* is restricted to northern states. In contrast, it is also widely distributed in the south. There are no reports on the availability of *P. manyar* and *P. megharhynchus* in the south. The present paper deals with some general aspects of ecology of baya birds with special reference to nesting by *P. Philippinus* in the Karnataka state.

This paper deals with the survey of various habitats for nesting colonies of baya, identification of different baya birds, orientation of nests, nesting behaviour, host plant selection and foraging activity.

Syslcmaiics

More than 95 species of weaver birds known to occur in the world, but India possess only four of them. They are classified as follows:

Class	:	Aves
Sub-class	:	Neornithes
Order	:	Passeriformes
Family	:	Ploceidae
Sub-family	:	Ploceinae
Genus	:	Ploceus
Species	:	*philippinus, bengalensis, manyar and megharhynchus.*

Common names

	Name of the Species	*Common name*
1.	*Ploceus philippinus*	Common Indian baya
2.	*Ploceus bengalensis*	Black throated baya
3.	*Ploceus manyar*	Streaked baya
4.	*Ploceus megharhynchus*	Finns baya

Keys for Identification

The key characteristics described in the book of Indian birds are as follows :

P. Philippinus

Male: During breeding season the crown is yellow, rest of the upper parts are dark brown streaked with yellow on the back. Breast yellow, throat and ear, dark brown and rest of the under part is cream buff (Fig. 32.7)

Female : Crown and back buff streaked with dark brown, neck yellowish buff, throat is whitish yellow, Breast is yellowish buff with brown streaks. During non-breeding season both male and female look more or less alike (Fig. 32.8).

Ploceus bengalensis

Male: During breeding season it has dull red crown or golden yellow crown, white throat and black band separating it trout the fulvous white underparts.

Female: Similar to other baya.

Ploceus manyar

Male: During breeding season it differs from other baya in having streak black on fulvous breast in both sexes. During non-breeding season both sexes look alike.

Ploceus megharhynchus:

Male: This bird was not noticed in the Western Ghats of Karnataka. During breeding season, its head and breast are bright yellow, and dark brown streaked on the above.

Female: Head and nape pale brown or yellowish brown with dark streaks. During non-breeding sexes are alike.

Among the three species recorded in our study area *P. philippinus* was most common bird and found in a variety of habitats and tree species. *P. bengalensis* was found nesting in grasses, reed beds and sugar cane fields and remained away from human settlements. The *P. manyar*, a single colony was found in a most protective habitat in the mid of wetland called Navile tank in the Shimoga city. The size is around 15 cm (6 inch) of the size of sparrow. Breeding season varies according to the monsoon season of the area.

Study Area

The study area is located in the vicinity of foothills of Western Ghats, receiving an average rainfall of 850 mm. The area is divided into three imaginary zones.

Fig. 32.7 Male baya in breeding dress waiting for females at helmet stage on the nest.

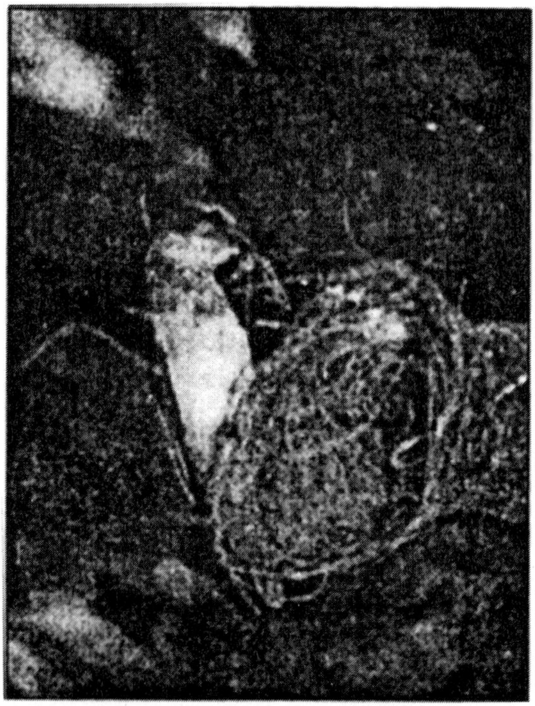

Fig. 32.8. Female baya with food material in the mouth sitting at the entrance chamber.

Zone I : Northern side

Zone II : North-east

Zone III : South-west

Each zone is spread in approximately 30 hectares land. The topography of the area include 13° 42' 00" N longitude and 75° 38' 20" E latitude. Usually the monsoon period commences in the month of June and closes by September.

Methods:

In the present study, we have enumerated only nesting colonies of *P. philippinus*. We have also encountered the colonies of other species like *bengalensis* and *manyar* species. The data on *philippinus* species is used for discussion in this paper. The colonies were located by following their songs and some time we have also taken the help of localities.

We have visited a total of 969 colonies in the year 1999-2000. As each male constructs 3-4 nests and each female pairs 3-4 times in a season, based on this particular reproductive behaviour, it is calculated that the total number of philippinus birds may be around 300-350.

Objectives

To know the following :

a. Orientation of nests

b. Nesting habitats

c. Host plants and their diversity

d. Breeding activity

e. Feeding activity

Orientation

Our study on orientation of nests by the baya P. philippinus revealed that the first year male or the yearlings without breeding dress and the sexually matured adult males have constructed their nests in all possible orientations. They are mainly lateral, central and peripheral sides of the host

Table 32.8. Orientation of nests by Ploceus Philippinus

No. of nests showing different types of lateral orientation with orientation diagrams.

Total No.	*E*	*NE*	*W*	*CNE*	*NEW*	*N*	*NW*	*SW*	*S*	*SE*	*NES*	*SN**	*No. of nests show-ing central orein-tation CN*	*No. of nests show-ing peri-pheral orien-tation*	*Total*
Zone I	40	13	7	6	4	6	5	- 5	4	2	3	4	12	10	121
Zone II	67	20	04	17	12	14	06	01 08	09	09	11	02	07	36	223
Zone III	39	16	04	10	08	09	08	- 08	06	05	04	04	04	05	130

E = East, W = West, N = North, S = South, C = Central colony, P = Peripheral, – = Direction of SW monsoon, O = Tree canopy = Nested part of canopy in respect of South West monsoon, % of lateral colony = 3.8, % of central colony = 93, peripheral = 3.0, * Bilateral.

plants and other combinations. The types of orientations studied in three zones are showed in Table 32.8.

In Zone I, lateral colonies were maximum (81.8%). Such colonies can face different directions in lateral positions. Among these again the colonies facing exactly eastern direction were maximum in number. In Zone II again lateral oriented colonies were maximum (80.7%). Among these the eastern side facing nests were maximum. A similar trend was also noticed in Zone III. There was a definite orientation pattern observed in these birds with respect to stem of the host plants. It is generally the entrance tube which directed towards outside while egg chamber was positioned towards the stem. Due to this particular orientation the entrance tube always remained oriented towards outside. The flying routes of the birds are so managed that the host stem or other vegetation do not pose any hindrance. Orientation of these nest was different in vegetated and non-vegetated areas, dug wells and irrigation channels, wherein they oriented in relation to their nearest surfaces. The *P. philippinus* hangs its nests according to the direction of prevailing wind.

Selection of Twigs

The selection of twig is also one of the important factors for nest weaving. Twigs should be generally thin pliable and pendent to horizontal branches (leaves in case of palms and grasses) are selected by the *P. philippinus*. Twig having thickness more than a thumb are usually avoided, possibly they can not be accommodated in the grip of claws during knitting of fibers in the nest initiation. Strongly upwarding branches are preferred. Twigs which are sufficiently tough to support the weight of the nest are chosen.

The onset of rainy season of the area is mainly known to influence the activity of the nesting by the baya birds. It operates for four months usually from June to September. Any variation in south-west monsoon is found to influence directly on the nesting behaviour and orientation pattern of the nests.

The comparative study of the three field zones in our study area revealed that maximum nest were hanged in the

East direction. The reason for construction of nest in the east may be to avoid or seek protection from South-west monsoon. But this pattern is not applicable in the case of the situations like open wells, steep wells, vegetation growing on vertical slopes and sloppy areas. This particular behaviour indicates the intellectual experience gained by the baya to seek protection from natural hazards (Frank Finn, 1990).

Nesting Patterns

A typical normal nest of *P. philippinus* is a bottle shaped nest called as normal nest and can be divided into three parts viz., a stalk, middle body and the entrance tube. The entrance tube is slightly shifted towards the egg chamber to form a bulging chamber (Fig 32.9).

Fig. 32.9. Normal nest of the Baya (*Ploceus philippinus*)

Abnormal Nests

Besides normal nests, various other types of abnormal nests were constructed by the sexually matured males and yearlings. Many variations can be seen in the nests of baya, which make the nest abnormal. The category of abnormality is in the structure of the nest or any part of it. Abnormal structure may appear due to deposition of parts, formations of additional parts, elaboration of parts and abolition of normal parts (Sharma, 1995). The second category of abnormality is regarding the position of nests or any part of it. The abnormality is classified in Table 32.9.

In our investigations in the Karnataka a total of 969 nests were visited from the three Zones I, II and III (Table 32.10). It is clear from the data that bistoreyed nesting was common and it was followed by multi-stalked nests. Why and how and under what situations such nests are constructed by the baya is an interesting and complicating question to be answered.

Though bistoreyed nests were common, four storeyed nests were also encountered in our study area. Multi-storeyed nests are noticed in those areas where comparatively long monsoon periods prevailed. In Karnataka, monsoon period is too short to fabricate 6-10 storeyed nests. Multi-stalked nests were also common in this area. Presence of two or more stalks provide extra attachment and strength to the nest which make a hanging nest prone to simple harmonic motions on windy days. To minimise the windy movements of the nests more than one support are used for stalking the nest. Additionally the length of the stalk is also kept short. It is mainly based on the habitat conditions. If the tip is harder than in order to have more security more than one stalks are weaved. Stalkless nests are also seen in the area and such nests are equally liked by the females.

The other type of abnormal nests are Bell Jar shaped, buttressed, blind or closed, chained, meshed, branched, fused and stomach shaped nests. In addition to this some normal nests possessed double openings in order to facilitate the quick visits by the mother.

The study on nesting pattern in the Western Ghats revealed that 11.9% were abnormal and out of this 14.03%

were multi-stalked and 40.35% bistoreyed respectively (Table 32.10).

Host Plants

It was observed that in our study area the common baya constructed its nests on 31 species of plants belonging to 18 families. The cocks constructed nest on various types of trees. Among all these trees most preferred species of plants were *Cocos nucifiera, Acacia planiformes* and *Pheonix sylvestris.*

A comparative study on three zones revealed that Zone I harbored highest number of normal nests (57%) followed by Zone III with 55.3% and Zone II (53.6%) normal nests respectively. The reason behind this type of variations may be due to existence of rich and diverse types of habitat resources and protection, the Zone I harbored maximum number of normal nests (Table 32.11).

Statistical analysis was applied by using Raunkiaers frequency test. It is apparent that low frequency values indicated higher number of normal nests with respect to individual species of plants and higher the frequency indicates low number of nests. According to Raunikiers frequency test the host plants like *Cocos nucifiera, Acacia planiformes* and *Pheonix sylvestris* showed lowest frequency number indicating maximum normal nests (Table 32.12).

Summary

1. The population of common baya was maximum in Zone II is due to rich and diverse type habitat and food resources.
2. More than 50% of the nests were directed towards the eastern side to avoid S.W. monsoon. In contrast the nest located on the banks of irrigation channels, open wells and sloppy areas showed different orientations.
3. Among abnormal nests bistoreyed abnormality was maximum and it was followed by multi-stalked nesting.

Table 32.9. Classification of Abnormal Nests

- Abnormal Nests
  - Simple Abnormal Nests
    - Monostoreyed Nests
      - Structural Abnormality (Abnormality due to abnormal structure of nest or any part of it)
        1. Bell-jar shaped nests
        2. Buttressed nests
        3. Blind nests
        4. Closed nests
        5. Stalkless nests
        6. Multistalked nests
        7. Chained nests
        8. Meshed nests
        9. Completed nests with double openings
      - Orientational Abnormality (Abnormality due to abnormal position of nest of any part of it)
        1. Fused or composite nests
        2. Branched nests
        3. Symmetrrical nests
        4. Stomach shaped nests
    - Multistoreyed nests (Abnormality due to abnormal position of nest or any part of it
      - Pseudo-bistoreyed nest
      - 2. Bistoreyed nests
      - 3. Chain-storeyed nests
  - Mixed Abnormal Nests

Table 32.10. Classification of Abnormal Nests

Zones	Year	Total No of nests observed	Simple abnormal nests																	
			Monostoreyed nests																	
			Structure Abnormality								Orientarional Abnormality					Pseudo-storeyed nests	Bistoreyed Nests	Chainstoreyed nests	Mixed abnormal nests	Total number of abnormal nests
			Bell Jar Shaped	Buttressed	Blind	Closed	Stalkless	Multistalked	Chained	Meshed	Fused	Branched	Symmerecial	Stomach shaped	Completed nests with Double Opening					
Zone I	1999-2000	374	2	5	-	-	4	6	4	2	1	-	-	1	-	2	18	2	-	47
Zone II	1999-2000	291		1	1	-	2	7	1	2	-	-	-	-	-	1	11	-	-	26
Zone III	1999-2000	294	1	2	1	1	7	3	4	-	-	-	-	2	1	1	17	2	-	41
	Total	959	3	8	2	1	13	16	9	4	1	0	0	3	1	4	46	4	0	114

Table 32.11. List of Plants and Bottle nests Recorded on host Plants

Zone	*Total No. of plants*	*Total nests*		*Percentage of normal nests*
		Normal (a)	*Abnormal (b)*	
Zone I	39	374	207	55.3%
Zone II	41	291	156	53.6%
Zone III	49	294	168	57.1%

Table 32.12. Frequency of Complete nests According to Raunkiaers Frequency Class

Zones	*Name of the Plant sp.*	*Frequency class*	*Density of Complete Nests*
Zone I	*Cocos nucifera*	A	8.6
	Acacia planiformes	A	17.0
	Pheonix sylvestris	A	42.0
Zone II	*Cocus nucifera*	A	8.6
	Acacia planiformes	A	12.5
Zone III	*Cocos nucifera*	A	9.4
	Acacia sp	A	10.0

4. For nesting 32 different plants species were selected, among those coconut, acacia and pheonix tree species were most preferred.
5. *P. philippinus* is not specific in hanging its nests, unlike other species of baya.
6. Population density of *philippinus* was maximum in the study area and it was followed by *bengalensis* species. A single colony with 14 nests of *manyar* species was also recorded.

Future Scope and Significance

Among the three species of baya birds, *P. manyar* may considered as endangered species in Karnataka. Before saying this, there is need to undertake a detailed survey in other parts of our country. There is a need to study the pest status of these baya birds in relation to different crops in different areas of our country.

h. Establishment of Orchidaria in the Western Ghats of Karnataka state

Preamble

Dr. T. Anand Rao is a senior orchid scientist of India. Based on his work the present chapter is prepared. The orchids are distributed every where in the Western Ghats. These produce fascinating beautiful flowers that mimic female bees, wasps and beetles. Some of them are attractive visually and olfactonly to insects and birds, which in turn visit them and try to copulate them. In this process of pseudo mating they apparently pollinate the flowers. The persons interested on these flowers collect them indiscriminately which has led to their virtual disappearance. There is a need to be conserved these orchids not only in the wild, but also in orchidaria for research and enjoyment of their ornamental beauty and medicinal importance. Accordingly an attempt is made to set up orchidaria corresponding to specific locations like Town Park, Kudremukh, Bhagavathi Nature Camp and Bannerghatta National Park. These are the three recognised hot spots of Orchids in the Western Ghats of the state. In orchidaria these may become commercially potential genotypes for propagation and trade.

Over the years in Karnataka apart from checklist of flora including orchids (Saldhana and Nicolson, 1976; Keshava-murthy and Yoganarshimhan, 1990; Anand Rao, 1998, 1999) there are comparatively less investigations made on orchids in their natural settings. Recently Rao (1998-99) has initiated a trend to explore recognised hot spots of orchids of Western Ghats of Karnataka and link it to orchidarium at Cauvery Nisargadham, Kodagu District. Now it is extended to the nature camp in Kudremukh National park and Town park Kudremukh. Further there is proposal to have a state level orchidarium at Bannerghatta National Park jointly supported by wild life wing of the Forest Department, Government of Karnataka and the Karnataka Iron Ore Company Ltd., Kudremukh. The main objective of establishment of orchidaria is to protect the germplasm of wild orchids, which are useful in many ways to our prosperity and further progress.

In Karnataka state one could see lack of organized orchidarium at state level holding wild orchids as germplasm stock under natural settings. Partially this lacuna has been met by establishing an orchidarium at Cauvery Nisargadhama, Kushalnagar, Kodagu District. To date, this orchidarium holds 85 species mostly collected from distinct vegetational types of the Karnataka part of the Western Ghats. Further two illustrated books on the orchids of Kodagu have been published (Rao, 1998,1999). Similarly, the orchid survey was conducted at the Kudremukh National Park during 1998-2000 to collect 59 orchids from distinct forest divisions. So far, this has yielded a few rare and unreported orchid species, which have been planted at the Bhagavathi Nature Camp orchidarium.

Kudremukh National Park in Karnataka is one of the best examples of Shola forests found in the Western Ghat belt of Karnataka and possess a rich assemblage of wild orchids of which a few of them are unreported to orchid flora of Karnataka. Further some of them previously thought to be restricted to a few areas in Karnataka state have been recorded in this chapter with a note on their occurrence in the Kudremukh National Park and also in other parts of Karnataka. The present collection in the Western Ghats has found to be rich and diverse in respect of orchids due to the fact that there are suitable niches for the guild system of orchids. The intensive collection includes unreported endemics which have been recorded earlier only in the states of Kerala and Tamil Nadu. They have a great potential also as earners of foreign exchange, besides having medical and commercial importance in the country itself.

Recently during the month of November, 2000 field work was undertaken to collect the seasonal orchids for planting both at Bhagavathi Nature Camp Orchidarium and also at Town Park, Kudremukh, under natural settings. To date, there are 25 orchids species planted at the above orchidarium which will be enriched in the course of the project. Ultimately there will be 120 species of orchids of medicinal, ornamental, mimic animations and highly scented orchids. At Bannerghatta National Park in a selected place under natural settings, 25 orchids have been planted during the month of November, 2000.

Culture of orchids in orchidaria

Epiphytic orchids are local specific but not host specific. In view of this, the point of consideration is how best they could be planted in the orchidarium. Among the wad orchids some of them thrive best in pots, wooden or bamboo baskets while others on trees under natural settings depending on the locations of the orchidarium. When pots are being used attention should be given to orchids growth forms, which fall under two types. Monopodial, namely elongation along the apical point of the unbranched axis or sympodial where new buds are produced laterally on the axis leading to branching system. Therefore when pot culture is practiced big pots with sufficient space should be used for sympodial orchids. When epiphytic orchids are grown in pots should be filled up with burnt brick of assorted sizes, chips of old bark, charcoal, bone chips and moss clumps. For ground orchids, decomposed cattle manure and organic manure with turfy loam forms a good mixture for rapid growth resulting in bright flowers. In earlier stages watering is necessary which in course of time should be kept in a shady place alternately subjecting the pots to sunlight and rain water during non-flowering seasons. Finally one should have or learn a sound knowledge of their growth features over the months by careful observations to have an ideal eye catching orchidarium. The orchids listed below were collected from Kudremukh National Park have been planted at Bhagavathi Nature camp, Kudremukh, Town park, Kudremukh and Bannerghatta National Park.

Orchidarium at the Town Park, Kudremukh and Bannerghatta National Park

1. *Aerides crispum* Lindl.
2. *Coelogyne breviscapa* lindl.
3. *Cymbidium aloeifolium* (L). Sw.
4. *Dendrobium barbatulum* Lindl.
5. *Dendrobium barbatulum* Lindl.
6. *Dendrobium macrostachyum* Lindl.
7. *Eria reticosa* Wt.
8. *Oberonia verticiliata* Wt.
9. *Rhynchostylis retusa* (L.) BL
10. *Cleisostoma tenuifolium* (L.) Garay
11. Trias *stocksii* Benth. ex. Hook.
12. *Luisia macrantha* Blatt. & McC.
13. *Cocologyne neroosa* A. Reich.
14. *Dendrobium aqueum* Lindl.
15. *Dendrobium ovatum* Lindl.
16. *Dendrobium crispa* Lindl.
17. *Vanda tessllata* (Roxb.) Hook ex Don.
18. *Eria microchilos* (Dalz.) Lindl.
19. *Acampe praemrosa.* (Roxb.). Blatt. & McC.
20. *Pholidota pallida* Lindl.
21. *Robiquetia josephiana* Manilal & Sathish.
22. *Liparis viridiflora* (BL.) Lindl.
23. *Cymbidium discolor* Lindl.
24. *Satyrium nepalense* Don.
25. *Taprobarea spathulata* (L) Christ.

Species present in the orchidarium of Kudremukh National Park, Bhagavathi Nature Camp, Kudremukh

1. *Aerides crispum* Lindl.
2. *Aphyllarchis montana* Reichb.
3. *Bulbophyllum proudlockii* (K & P)
4. *Calnathe triplicate* (Willem) Ames
5. *Coelogyne breviscapa* Lindl
6. *Cymbidium aloeifolium* (L) Sw
7. *Dendrobium lawianum* Lindl.
8. *Dendrobium barbatulum* Lindl.
9. *Dendrobium macrostachyum* Lindl.
10. *Eria microchilos* Wt.
11. *Eria Retieosa* Wt.
12. *Habeneria longicorniculata* Crash. (Ground orchid)
13. *Ma/axis versicolor rheedi* Sw. (Ground Orchid)
14. *Habeneria marginata* Coleb. (Ground orchid)
15. *Nervilia crociformis* Zoll. and Mor. (Ground Orchid)
16. *Oberoma verticillata* Wt.
17. *Peristylus gooderoides* Lindl.
18. *Porpax reticulata* Lindl.
19. *Rhynchostylis retusa* BL.
20. *Sarcanthus pauciflorus* Wt.
21. *Trios stocksn* (Benth) Hook. F.
22. *Luisia birchea* B.P.
23. *Luisia macrantha* Blatt & McC.
24. *Coelogyne nervosa* A. Reich
25. *Epipogium roscitm* (D. Don) Lindl.
26. *Dendrobium aqueum* Lindl.
27. *Nervilia argona* Grandl. (Ground orchid)
28. *Dendrobium ovatum* Lindl.

29. *Dendrobium crispa* Lindl.
30. *Vanda tessellata* (Roxb) Hook.
31. *Dendrobium erepidatum* Lindl.
32. *Gastrochilus acaulis* Lindl.
33. *Luisia zeylanic* Lindl.
34. *Acanthe philippensis* bicolor Lindl.
35. *Malaxis rheedei* Sw. (Ground orchid)
36. *Trios sfocJcsii* Benth ex-Hooker.
37. *Acampe praemrosa* (Roxb) Blatt & McC.
38. *Gastrochilus flabliformis* (Blkatt & McC). Saldhana
39. *Luisia birchea*
40. *Rhynchostylis retusa* Bl.
41. *Dendrobium soni* (Hybrid)
42. *Pholidota pallida* Lindl.
43. *Chiloschrista glandulosa* Schlter.
44. *Liparis viridifolia* Lindl.
45. *Diplocentrum congestum* Lindl.
46. *Tainia bicornis* (Lindl) Reich.
47. *Rhynostylis retusa* (L) Bl.
48. *Cymbidium discolor* Lindl.
49. *Satyrium nepalense* DON.
50. *Luisia macrantha* Blatt & McC.

i. Habitat Degradation for Peafowls

Dr. S. Basavarajappa, Lecturer, Department of Applied Zoology, Kuvempu University, B.R. Project has studied the degradation of habitat conditions for the Indian peafowl *Pavo cristatus* in the Channagiri Taluka of Karnataka. The information availed from him is used in this topic.

Peafowls are a source of beauty, joy and recreational pleasure to mankind forever. Being a national bird of India, it is considered as a potential species for its ecological, economical, ethical and medicinal values (Ali and Ripley, 1983; Ensminger, 1992). Peafowls are large bodied gallinaceous (order: Galliformes Family: Fasinidae) birds inhabit diversified agro-climatic habitats of India. Since time immemorial, it has been known that India was the cradle for the evolution and distribution of peafowls. Peafowls are native to India, Java, Malaya and Congo (Ensminger, 1992). Several species of peafowls and their sub-species are found in various parts of the globe. Indian blue, Java green and Congo peafowls are most common in occurrence. Their distribution is shown in Table 32.13.

Table 32.13. Different species of Peafowls and Their Distribution

Sl. No.	*Common name*	*Scientific name*	*Place of occurence*
1.	Indian blue or Crested peacock	*Pavo cristatus*	India
2.	Irish peacock	*Pavo bicalcratus*	Irish
3.	Tibetian peacock	*Pavo thibetanus*	Irish
4.	Japan peacock	*Pavo muticus*	Japan
5.	Green peafowl	*Parus caudatus*	Japan
6.	Javanese peacock	*Pavo spiciferus*	Sumatra
7.		*Pavo javanicus*	Islands cf Java

Despite suitable habitat conditions, the peafowls are facing severe threat under different agro-ecological habitat conditions of Karnataka. Though the species is protected legally, it has been poached illegally. As an author, I am worried to note that in many cities of Karnataka peafowls products, made up of its feather (Fan) are sold in the market illegally. Due to illegal poaching and habitat loss, in the near future the peafowls may be pushed into the category of endangered birds. In order to evaluate its population density under changed environmental conditions in Karnataka the present study is undertaken.

Study Area

Physiographically the study area, Channagiri is a part of Davanagere district occupies a central position in the Karnataka State. It constitute the southern part of Deccan Peninsula, lies in between 13°27' to 14°39' N latitude and 74°38' to 76°40' E longitude of 671 meters above msl. The landscape consists of vast stretches of planes with scattered small hillocks. The natural vegetation is sparse, more often diversified with deciduous plant species at the valleys of small and uneven hillocks and croplands. Details about the habitat is explained in Table 32.14.

Table 32.14. Habitat Features of Indian Peafowl *Pavo cristatus* in Channagiri

Habitat	*Physical conditions of land*	*Major crops nearby*	*Dominat vegetation*
I.	Plain surface with Irrigation, intensively cultivable (low land) land	Paddy	Scattered horticultural gardens viz., Areca, Coconut, Banana and Mango plantation.
II.	Uneven land surface with rainfed cultivable land small to medium hillocks nearby Santhisagar lake	Jawar, Maize	Small patch of scruby and Scattered vegetation with horticultural gardens.
III.	Small to medium hillocks uncultivable (upland) land.	Small Patch of jawar & maize crops	Deciduous forest, dominated by teak bamboo sp.

The observation was made during the period between 1998 to 2000 in different agro-climatic ecosystems of Channagiri Taluka of Davanagere District. Three different habitats were identified at random viz., (I) Plain area of 966.81 hectares, (II) Uneven and hilly area with sparse vegetation of 300 hectares (near Shanthisagar—a biggest lake in south India) and (III) Hillocks with dry deciduous vegetation of 600 hectares (Joldhal village).

Observations

Variations in the population densities of peafowl in the above said areas are printed in Table 32.15. It is clear from the observation that the Study area III near Joldhal supported maximum number of pear fowls (190-195) followed by 103-125 near Shanthisagar (Zone II) and only 65 to 96 in Zone III respectively (Table 32.15). On an average cocks were 42.3, hens 192.7 and juveniles 148.3 in the entire area.

Table 32.15. Peafowl (*Pavo cristatus*) Population at different Habitats of Karnataka

Year	*Individuals*	*Habitat I*	*Habitat II*	*Habitat III*	*Total*
1998	Peacock	12	16	20	48
	Peahen	45	68	8	121
	Juvenniles	39	41	87	167
	Total	**96**	**125**	**115**	**336**
1999	Peacock	08	12	20	40
	Peahen	38	65	89	192
	Juvenniles	25	35	84	144
	Total	**71**	**112**	**193**	**376**
2000	Peacock	08	12	19	39
	Peahen	38	59	88	185
	Juvenniles	19	32	83	134
	Total	**65**	**103**	**190**	**358**
Total	Peacock	28	40	59	127
		(9.3)	(13.3)	(19.7)	(42.3)
	Peahen	121	192	265	578
	(Mean)	(40.3)	(64.0)	(88.3)	(192.6)
	Juvenniles	83	108	254	445
		(27.7)	(36.0)	(84.7)	(148.4)

Note: Each value is a mean of 24 observations.

In general it is evident from the studies on the population density that the number declined from 1998 to 2000. It was a total of 416 birds in the year 1998, it declined to a total of 376 in 1999 and further declined to 358 in 2000 respectively. This particular observation revealed that there are some factors operating in the study area which are responsible for the decimation of peafowls. The major reason for the decline in birds could be habitat loss and shrinking of breeding areas during breeding seasons.

Table 32.16. Commonly Used Roosting Plant Species of Peafowl *P. cristatus* in Karnataka

Habitat	*Common name*	*Scientific name*	*Family*
I.	1. Mango	*Mangifera indica*	Anacardiaceae
	2.Coconut	*Cocos nucifera*	Arecaceae
	3. Niligiri	*Eucalyptus* sp.	Myrtaceae
	4. Jali	*Acacia* sp.	Mimosaceae
	5. Syzygium	*Syzigium* sp.	Myrtaceae
	6. Tamarind	*Tamarindus indica*	Caesalpiniaceae
II.	1. Mango	*Mangifera indica*	Anacardiaceae
	2. Niligiri	*Eucalyptus* sp.	Myrtaceae
	3. Rosewood	*Dalbergia* sp.	Fabaceae
	4. Jali	*Acacia* sp.	Mimosaceae
	5. Neem	*Azadirachta indica*	Meliaceae
	6. Tamarind	*Tamarindus indica*	Caesalpiniaceae
III.	1. Srigandha	*Santalum album*	Santalaceae
	2. Syzygium	*Syzigium* sp.	Myrtaceae
	3. Niligeri	*Eucalyptus* sp.	Myrtaceae
	4. Rosewood	*Dalbergia* sp.	Fabaceae
	5. Teak wood	*Tectona grandis*	Fabaceae
	6. Terminalia	*Jerminalia* sp.	Fabaceae
	7. Bamboo	*Bambous* sp.	Fabaceae

It is observed that the natural grass are replaced by agricultural crops. Any alteration or destruction in natural habitat will prove a threat to most desired species and favourable to a few other species. Additionally the activity of overgrazing by livestock, deforestion and encroachment of land for agriculture are collectively affected the survival of peafowls in this area.

Vegetation

About twelve species of trees belonging to eight families are used by the fowls for roosting (Table 32.16). Amongst the tree population, the members of *Anacardiceae* and *Myrtaceae* are most preferred species. It is observed that peafowls choose fruit yielding tree species also. It is found that peafowls were more selective while roosting.

Summary

Though the species still exists in large population and is not yet endangered but it is a rapidly depleting species. It is protected legally, but still illegal activities are on. The bird is killed for its meat and feathers. Additionally the changes in its habitat conditions are also responsible for the loss of breeding grounds thereby decrease in the peafowl population. There is an urgent need to survey the population of peafowls in the entire country so that its wild life status can be adjudicated properly.

1. Importance of Bats

Endemic Bat (Otomops wroughtonii)

Being born in Belgaum district, I was much attracted to read and know more about the endemic and rare bats found only in the caves of Talewadi village of Belgaum district. *O. wroughtonii* is considered as one of the rarest insectivorous bats in the world. It is endemic to Belgaum district and not found anywhere on the earth. The cave is large dense with large number of millipedes and some other insects. The rocks in the caves have crevices in which the bats reside. Fecal dropping are comparatively less in this area compared to other types of bats. It is very difficult to locate these bats because they fly very swiftly. Early reports by Prater (1960) revealed that these bats lived in groups and he killed around 40 of them. It is believed that the construction of dams and mining activities may not harm the bats in the future. But how far it is true and this can adjudicated only by the bat biologists.

Painted bats

The painted bat *Kerivoula picta* is widely distributed in India, there are very few sight records of this bat from Maharashtra. Systematic studies on this bat is not undertaken in India. Banana plantation is the prime habitat for this bat species. These bats are facing a serious threat due to the clearing of banana plants to make a way for Sapota orchard (*Achras zopota*).

Fruit Bats

Bats belong to order Chiroptera which can be further sub-divided into mega and micro-chiropterans. Mega-chiropteran bats are confined to old world tropics and are exclusively phytophagus whereas micro-chiropteran bats are mostly insectivorous. In Indian sub-continent out of the 12 species of fruit bats three species are most commonly known. Those are the flying fox (*Ptcropus giganteus*), the fulvous fruit bat (*Rousettus leschenavlti*) and short nosed fruit bat (*Cynopterus sphinx*). About 70-100 species on insect eating bats species are found in different parts of India. Their diversity is more in the Eastern and Western Ghats, Himalayas, Andaman and Nicobar Islands.

The frugivorous bats and their pest control measures revealed that these are eco-friendly rather than thinking them as pest on orchards. Approximately, 250 species of bats of the paleo-tropical family, Pteropodidae and the neotropical family, Phyllostomidde, regularly visit plants for either fleshy fruits of floral resources. The foraging activities of a frugivorous bats can be simply described on the following way: At dusk fruit bats level their roosting site and commit to a particular resource patch which is often fruiting tree. This bat then selects a fruit from those available and either eats it or carries it down to its roosting place. During feeding it might feed on an entire fruit or sometimes a part of it leaving the remains as masticated or rejected pellets or uneaten parts, usually found under the roosting site. The bat may carry a gut load during its foraging activities. At dawn, bats return to their roosts and a final gut load is either eaten or regurgitated.

In the eyes of a frugivorous bat, not all plant species are equally attractive. Within and between fruit trees, there is considerable variation in ripeness, size and other physical or chemical attributes. Fruits eaten by bats are usually fleshy or have a musty odor and high sugar content. The role played by bats as seed dispersers help in promoting gene flow, making seeds less obvious to seed predators and moving seeds to more favorable sites for colonizing. In case of pollination, the benefit the plant receives when pollinated, depends on the amount of pollen deposited. A large amount of pollen increases the proportion of ovules that could develop. The second possible benefit of depositing is that competition among pollen grains may increase the quality of progeny in plants. Flower size, the number of flowers visited and the amount of pollen produced by the flower, influences the quality of pollen that the bat carries.

Movement of pollen between individuals thus promotes plants to produce smaller amount of nectar so that bats would visit more than one individual flower to satisfy foraging requirements. Thus in case of keystone species such as Ficus spp, bats help in the propagation of seeds, promote dispersion of seeds to various patches far away from the parent plant. Thus, there exists a kind of mutual relationship between fruit bats and the plants served by them for pollination and dispersal.

Most bats are seasonal at various patches. The phenologies of fruiting correspond directly to the reproductive status of these bats. As frugivores, both megabats and man have similar tastes and both share many fruits. So man through the tending of wild fruit trees and planting to enhances the chances of experiencing heavy damage to fruits in farm lands and orchards. One possible reason for the increasing menace of bats in fruit orchards is the dwindling population of wild fruit trees and a general reduction of forest cover. The rapid loss of the ficus trees has led bats to forage in orchards is perhaps a strong factor. Thus certain management techniques could be considered in managing this menace."

The techniques can be classified into two main categories:

(a) The roost management and (b) Crop management.

(a) Roost management

1. Physical methods: Making of sounds at their roost site which are present in window caves or palm trees. The making of loud sounds could keep bats away for some time.

2. Chemical repellents; chemicals having a bad odour drive them away. This should be sprayed when bats leave their roost during the late evening hours. For example fresh cow dung mixed with water or goat dung or paint.

(b) Crop management

The success of crop protection methods depends on the availability of alternative food resources in the nearby vicinity. Roost control methods are effective only if alternative foraging habitats are available closely otherwise they would continue to forage in orchards although their roosts may be shifted.

Plantation strategies

The long-term strategies include the following: Buffering with other fruit crops, planting of other fruit trees with higher sugar content could alleviate the pressure. Block plantation divide the plantation into smaller compartments which can be covered by nylon or mosquito nets, conservation and planting species such as Ficus bengalensis, Ficus glomerata, Ficus religiosa, Ficus benjamina and Ficus mysorensis, would certainly reduce the load of bats in orchards and farmlands. Methods suggested above are at the best speculative since a great deal yet has to be known about behavior and the type of destruction caused. While farmers could experiment with these methods research on some of the fundamental issues should be done simultaneously future studies should focus on bat plant range of frugivorous bats and their exploitations in managing bat damage to commercial fruits orchards but management must focus on regulating fruit bat population to tolerable levels rather than their elimination. This could cause cascading effects on both natural and man-made ecosystem.

All Fruit Bats are not Nuisance

The suborder megachiroptera consists of 175 species of fruit hats. Apart from food habits and size the mega bats differ from micro bats in having a small claw on the second digit of lore limbs. Fruit bats have relatively large eyes, and are well adopted for nocturnal vision. The olfactory system plays an important role in the location of food. The sensitive nose is useful in detecting the aromatic scent of ripe fruits in darkness. The fruit bats play a beneficial role in pollinating flowers and dispersing seeds. They do damage commercial fruits while visiting orchards for foraging. There is a necessary to investigate the quantification of damage caused by fruit bats. However, the Wild Life Protection Act 1972 has already classified bats as vermin.

K. The Indian Sarus Crane (Grus antigone)

The Indian Sarus crane, (Grus antigone) is distributed mainly in the northern part of India. Gole (1989) recently carried out a survey to estimate the distribution of the species in the range mentioned by Ali and Ripley (1969), though it is a wetland species, it also developed flexibility in its habitat requirements. Ecology of the sarus crane is not studied extensively and the previous work includes qualitative description, their distribution, diet, habitat, some aspects of its morphology and eggs. Among the four species of cranes reported from the Indian continent, the sarus crane is resident of Koeldeo National Park and the rest are migratory. Among the migratory cranes, Siberian crane was a permanent visitor to India, which is becoming rare during the past 4-5 years. The Common Crane and Demoiselle crane are occasional visitors. These are regularly sighted in the Western Ghat region of Karnataka state. These frequently visit the wetlands, and irrigated croplands of Shimoga district, Hidakal reservoir of Belgaum, Haveri district and other water tanks.

An in-depth study on the Sarus crane was carried out by Ramachandran and Vijayan during 1984 to 1988 at Koeldeo National Park located 2 km away from Bharatapur city in Rajasthan on the south-west of Mathura and 50 km away from Agra city. The total area of Koeldeo Park is 29 sq. km, out of which 8.5 km are inundated and the area dries up during summer leaving some pools in deeper areas. The Park is surrounded by agricultural lands and settlements. Some of the area forms good grasslands and the rest is deciduous thorny scrub jungle.

Population and Distribution

The population of the Sarus crane remain almost same throughout all parts of the year except during spring and summer. During summer most of the birds leave the park for feeding outside, while some pairs remain inside throughout the year as resident group. These may be breeding pairs of the park. During breeding period most of the cranes are confined to the waterspread areas of the Park. In 1986 with the onset of monsoon and subsequent flooding of the Park, the massive summer congregation of Sarus dispersed leaving only around 24-25 birds inside the park which were breeding in the park. The observation on breeding population of Sarus in the park during the period from 1969 to 1988 is shown in Table 32.17.

Table 32.17. Breeding Population of Sarus Crane in the Koeldeo National Park

Year	*No. Pairs*	*Density/sq. km*	*Source*
1969	27	0.9	Walkinshaw (1973)
1980	16	0.56	Belinda Breedery
1982	14	0.49	Ali and Vijayan (1986)
1984	3	0.11	Natarajan
1985	11	0.38	Natarajan
1986	10	0.34	Natarajan (1994)

Breeding Season

The main breeding season during 1982-1985 was August to October. But there are reports that, it can also breed starting from June to October. The breeding season of Sarus is highly variable, it may stretch upto to March.

Nesting Site

Usually the nest is constructed on a slightly raised area in the submerged parts of the wetland. The nesting site is invariably surrounded by water, the depth of which varied between 15 cm to 20 metres. Nine out of eleven nests in the year 1985 were on thick grass patches and two were on Acacia mounds. The nesting site was selected in such a way that one or two sides of the nest have open expansions of grasses which may help in detecting the predators at a distance.

The nesting material varied from nest to nest and was dependent on the nesting site. They seldom bring materials to the nest and construct it. Of the eight nests intensively studied by Ramachandran and Vijayan (1992). The nest was usually a pile of grass on a mat of Paspalum distichitm with nest radius of 1 to 1.5 metres. The base of the nest is broader with a slight depression on the top for the eggs. The grass pile reaches 70 cm in height from the surface of the water. It may also be constructed by Acacia twigs and the Paspalum grass. Both male and female participate in the act of nest building for period of one month. Repair and strengthening of the nest takes place even after egg laying. There was a gradual reduction in the breeding population of the bird in the Park during the past two decades.

Territoriality

These birds maintain a well established territory during breeding. The pair remain in the same area throughout the year except in the summer months. However, during non-breeding season they cover large area for foraging. The breeding territory ranged around 0.068 sq. km and the nest is built on one corner of the territory. Of the eight nests studied only one was in the centre. Most of the territories had two zones, a shallow water area where parent birds feed and a deeper zone in the immediate vicinity of the nest. The deeper area is reserved for the young ones. By the time young ones hatch out, the water recedes the area becomes shallower and all the birds feed in that area.

The incubation period for the bird is 28 to 34 days. Both sexes participate in incubation. In the initial stage of incubation, while one parent incubating the other remained nearby and engaged in the strengthening the nest. It is observed that during exchange of duty of incubation the partners give duet dance or prolonged unison call and then go for incubation. It is believed that such display will strengthen the pair bonding between the two. Usually male spend more time in incubation than female. The chicks begin to move along with parent on the day of hatching itself. The parent carry the food in their beak and feed the chicks. The chick always remain with the parent for one season then move away to live independently.

Food and Feeding Habit

The food material observed include rhizomes of *Nymphaea* sp., *Scirpus tuberosiis, Eleocharis plantaginca,* insects, grass seeds, mollusks and snakes. During summer when they move out they feed on wheat seeds from the croplands in dry areas. These bird roost in pairs or with family parties. The roosting site will be usually on grass patches of the wetlands. Sometime they roost along with Siberian cranes. Since these Sarus cranes use moderate and heavily flooded grassland for feeding during breeding season, these habitats have become a critical sites for the survival of these species.

Annual Cycle of Sarus Crane

The annual biological cycle of the Sarus cane is closely interwoven with the seasonality of the climatic regime. Since the Koeldeo National Park is the last safe refuse for these cranes in India, they aggregate in large numbers near the pools of the park. With the onset of monsoon and subsequent flooding the pools in the park, most of the Sarus cranes disperse leaving around 25-30 birds in the Sanctuary, which forms the breeding or resident population of the park. They are territorial during this season and all the chicks will be out by the end of October or November, if conditions are favourable then during January and February a few pairs breed and produce chicks again. Once the water pools dry cranes from neighbouring areas congregate and making them non territorial.

Koeldeo National Park is an excellent example of an insular habitat separated from surrounding areas, which are agricultural lands and human settlements. Due to heavy anthopogenic pressure around the park, it revealed that during the last two decades there was a declining trend in their population in the park. Another threat to the birds is contamination of water by industrial effluents and pesticides residues into the park waters. There is a need to develop the conservation strategy for Sarus crane by considering the above discussed factors.

l. Southern Crow Pheasant Centropus sinensis parrotius Stresemann

Relatively less information is available on the crow pheasants of India. Little information is available on the biology of Indian crow Pheasant (*Centropus sinensis sinensis*). The detailed study on Southern pheasant (*Centropus sinensis parrotius*) was carried out in Point Calimere Wildlife Sanctuary, in Tamil Nadu, by Dr. V. Natarajan in 1988. He has studied six different breeding territories in Kodikkanal village. Sightings of breeding pairs were marked on them and the territories are calculated. During the season of 1986 to 1989 study was carried out and a total of 34 nests were located. The study area Gana Koeldeo National Park is shown in Fig. 32.10. Among these, 21 nests in the village area were continuously monitored and the clutch size, incubation, hatching rate and breeding success were recorded.

Breeding Season

At point calimere the crow pheasant raises a single brood in a season which extends from November to May. The egg laying was found during December to March.

Territory

The breeding pair occupy a well-defined area defended by both partners. The size of territory usually varies between 0.9 to 7.2 ha with a mean of 3.8 ha. In crow pheasant nests may be seen in same tree for consecutive years. Other pairs entering the territory were threatened by making a sound K-wiss call. Other birds like crow, shikra, brahminy kite were tolerated in the territory but not on the nesting tree. The coop-coop-coop-call by the pairs is an advertise to threaten the intruders and keep them away and this call is heard frequently during the breeding season.

Courtship Behaviour

The display leads to mating. For the first time the male arrives with a frog in the mouth in its bill and hop along with the female on the branches of a tree. The female then glide from the tree to the ground and run for a short distance with wings open, vibrating, uttering a harsh cry sk-e-e-aw, similar to the call of common crow pheasant. Both fly on a tree and hop again from branch to branch, they copulated two to three times. The frog in the mouth will either eaten by the male alone or it will be shared with the female also.

While searching for a nest-site the crow pheasants hop from branch to branch, sit on a perch and produce calls, each partner respond to the calls. Both sexes usually examine the site and take part in nest building. Dhindsa and Toor (1981) reported that the common crow pheasant (*Centropus nnensis sinensis*) completed its nest within three days. In the case of Southern crow pheasant it required 7 days and the eggs were laid before completion of the nest building. Crow pheasant prefer thorny plants to build their nests. In Karnataka the common crow pheasant usually construct its nest in thorny lemon trees. These birds construct their nest like a globe or dome with a lateral entrance. Some time it may build nests deep cup like.

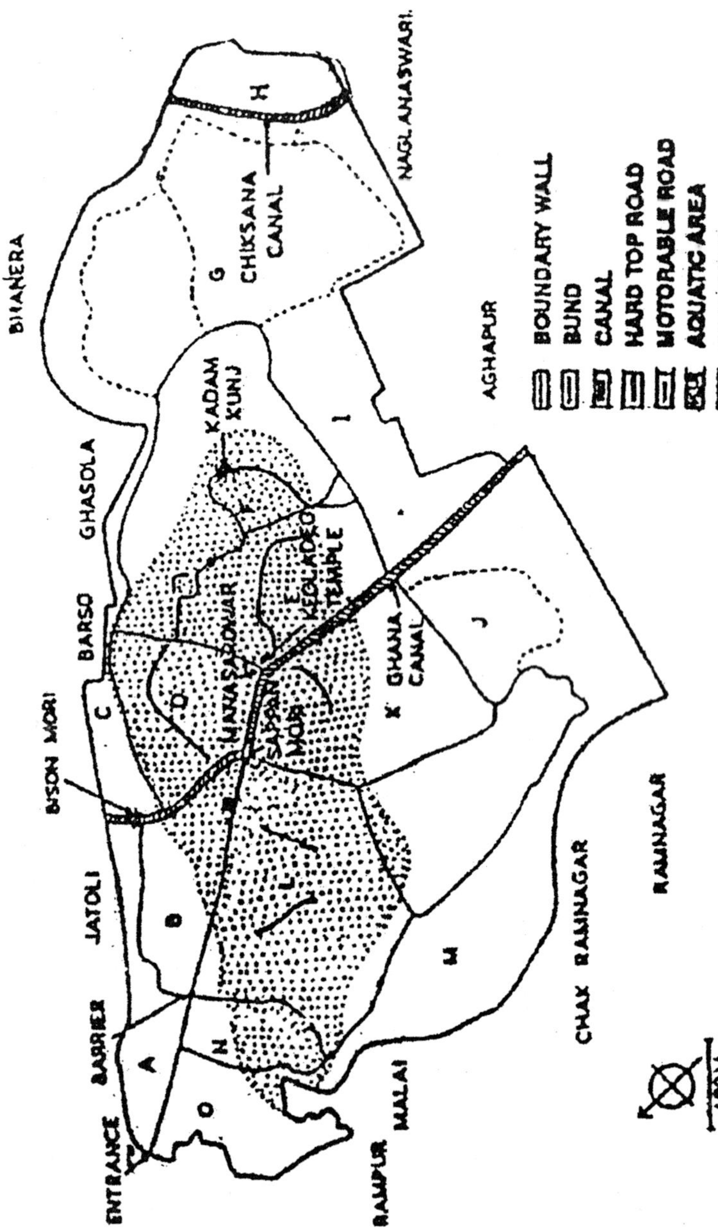

Fig. 32.10. Map of Keoladeo National Park

Egg Laying, Clutch Size and Incubation

Usually eggs were laid at the time of completion or about to complete the nest. Afterwards both the parents mostly become busy in completion of the dome. The eggs were oval in shape and chalky white in colour, become yellow stained as incubation proceeds. The length of the eggs was 34.8 mm and breadth 14.8 mm was respectively. The clutch size varied from one to four eggs. Both sexes take part in incubation during day time but at night time only one will incubate the eggs. During first three-four day old young ones are fed with regurgitated food, especially snails. Both parents feed the nestlings. The number of feeding visits increased till 16th day declined thereafter. Parents visited the nest 4-5 times per hour. Both parents involved in cleaning the nests.

Table 32.18. Plumage Changes In Crow-Pheasant Nestlings

Age in days	*Plumage*	*Remarks*
1	Body colour dark	Total body length 76 mm
2 and 3	Remiges and rectrices feather pins emerge	
4	Feather pins emerge in alula, upper wing and upper tail coverts	Eyes started opening
5	Spinal, humeral, capita tract pins beginning to emerge	Bill black, Eyes half to three quarters open
6	Eyes fully opened	
7	Ventral tract had two longitudinal rows of feather pins emerging	Infection by the tick Haemaphysalis intermedia was noticed on the wings
8 to 12	Feather pins of primaries ready to open	Tick infection noticed on dorsal side of the body
13 and 14	Feather pins of few primaries and wing coverts emerge from sheath	Ticks were noticed under tail coverts
15 and 16	All remiges and rectrices feather pins emerge from sheath. Spinal, capital tract pins appear as black brush-like feathers	
17 and 18	Plumage well developed. Nestlings fledged on 18th day.	

Growth of Nestlings

The just hatched nestling will be black in colour with eyes closed. Dorsally covered with long white hair like down feathers named as trichoptiles by Friedmann (1930). The upper mandible is black with pink edges. The center of the belly is pinkish, the legs and claws grayish in colour and the plumage changes with age (Table 32.18).

The nestlings were known to exhibit, escaping through the rear end of the nest chamber, excretion of foul smelling obnoxious sticky fluid and hissing sound like a snake when they are threatened by a outsider was observed.

Fledging

It was known to occur in about 19-21 days. Even after fledging the young were known stay near the nesting tree for few more days and later they slowly move to adjacent trees by hopping along the other branches. The parents continue to feed them even after fledging atleast for about one month.

The ringing experiments conducted by Natarajan (1996) revealed that even two months after fledging the young ones were seen living and feeding with their parents and dispersed thereafter.

The predators of the eggs include jungle crow, house crow, brahminy kite and the common mongoose *Herpestis edwardsi*. These predators will be chased away by the parents when they come near the nest.

m. White Winged Wood Duck, *Carina scutulata* in India

It is one of the most endangered ducks of India. It was sighted commonly in the north-eastern parts of India till the beginning of twentieth century. Later on there has been a steady decline in the population of this duck. In Dibrugarh and Lakhimur districts of Assam, its number declined from an estimated 1900 to a minimum of 44 by 1992 (Green, 1992, 1993). Green has documented in detail about the distribution of this critically endangered bird population. Dr. H. S. A. Yahya of Aligarh Muslim University carried out a pilot survey to collect on hand information on the status and habitat conditions of this duck in five key habitats of Assam and Arunachal Pradesh in 1993.

His results revealed that out of 65 birds estimated in the India, Dr. Yahya sighted only 26 animals in intensive survey of about 40 days. More number of these birds were located in Doom Dooma and Dibru Saikhowa Wildlife Sanctuary. About three key sites were surveyed in Arunachal Pradesh and no bird was sighted except their calls and sound. The estimated population number 65 birds may be an underestimate. According to Yahya the total number of white winged wood ducks may be around 100 pairs in the India.

The habitats of the duck are severely depleted and no site in the north is free from human influence. Depletion and fragmentation of habitat and poaching of young/eggs appears to be the main reason for the drastic decline in the population of this bird. Poaching and selling of adult birds still continued in these key areas. Several such cases were noted by Dr. Yahya of Aligarh. Arunachal beel is a suitable habitat for these birds but the quarrying, selective logging and hunting by the locals are common problems to be encountered by these animals.

Conservation Strategies Required to be Adopted

a. There is an urgent need to protect the habitat from further degradation. A sort of fencing is required at vulnerable sites like Ubahata, Littong and Namholong.

b. There is a need to locate viable nesting pairs and protect them from poaching. Artificial nest boxes may be supplied and monitored for these birds.

c. More staff and manpower is required to check and limit all illegal activities in their habitats. The Littong and Kakopathar forest areas may be announced as a Wildlife sanctuary and may be given special protection. Updating the present captive breeding centers is needed.

d. There is also an urgent need for creating awareness among the public about the wildlife status of the bird.

n. Conservation of Indian Marine Turtles

Turtles are referred as terrestrial animals based on their place of origin and birth, however they spend most of their life time in the water, except that they come to shores during breeding season. Turtles are given much importance in terms of their study and conservation as well. Marine turtle population is declining day by day due to rapacious turtling and over exploitation. For the past few decades their survival is threatened with extinction and they are under constant anthopogenic pressure. The marine turtle flesh has been an item of tasty table delicacy since long. As a fellow harmless creature both in water and on land, they posed never any threat either to marine ecosystems or to human beings. These poor creatures are troubled by man directly or indirectly through ages.

The marine biodiversity of turtles is in peril and their conservation is at stake. There is an urgent need to protect them from extinction. Conservation scenario of world wide sea turtles in general and Indian sea turtle in particular is full of frustration. Major threats to sea turtles include mortality caused by exploitation of turtles and their eggs (Abraham, 1989,1990 ; Yadav, 2000). Industrialization, human settlements and other types of habitat alterations causing serious concern in the protection of turtles on one hand and over-exploitation for their protein rich meat and money making greed through trade on the other. This menace is taking place not only in India, but also throughout the world (Silas, 1984a, 1984b).

The sea turtles are slaughtered on the beaches, the eggs are raided by the birds, animals and humans, thereby checking the chances of their progeny to continue (Pritchard, 1980). Indians by and large have been appreciably ethical towards all types of diversity. The marine turtles should be viewed as friends of human. Out of the seven worldwide species of sea turtles the following species are found in Indian maritime.

1. Olive ridley *Lepidochelys olivacea*
2. Green turtle *Chelonia mydas*
3. Leatherback turtle *Dermochdys coriacea*
4. Hawksbill turtle *Eretemochelys imbricata*
5. Loggerhead turtle *Caretta carctta*

The turtles have been given their due place in the Hindu mythology by recognizing them as second incarnation of lord Vishnu calling as Koorm Avtar. They are the important members of our biodiversity due to their biological, socio-economic, educational and scientific values as natural resources along the national coastline of about 7500 km. Though their population is dwindling, none of them are called as endemic and these have been placed in Schedule 1 of the WPA 1972 and listed in the Appendix I of the CITES. Due to their peculiar way of life their protection has become difficult. The marine turtles being a migratory species recognizes no boundary.

Conservation

Fortunately the Indian coasts have been supported by a number of sea turtles as their feeding and breeding grounds extending from Bay of Bengal to Andaman Nicobar islands, Gulf of Mannar, Indian Ocean, Arabian Sea upto Gujarat Coast. With reference to these animals the World's biggest rookery is Gahirmatha and Bhitarkanika in Orissa coast should be worth to mention. In recent times these resting beaches have become serious concern both at national and international levels. The Indian interest in sea turtles is not as old as other countries, who have spared more time and money for conservation and management of these animals. The scientific community should be concerned about the habitat disturbance, mass killing and over exploitation of these poor creatures. The conservation strategies undertaken under the leadership of Dr. Silas (1984), CMFRI, Cochin is landmark in the management history of sea turtles in India. It has been resulted in the great awareness in the scientific community and public as well. There is need to protect apart from ridley turtle, other turtles along the coast of Andhra Pradesh, Tamil Nadu, Karnataka,

Goa, Maharashtra and Gujarat states. Presently the Orissa having Olive Ridley as their prime land gained special momentum. This place is facing a number of human induced threats like mortality, predation, illumination, over exploitation and domestic and international trade.

Problems affecting the turtles

The Indian turtles are facing manyfold problems and some of the recent problems are listed by Yadav (1999) as follows :

1. Anthropogenic activities are affecting the habitat loss, inundation and poaching of eggs.
2. Nesting places in the beaches are threatened by resort developers and illegal human settlements.
3. Mechanized trawl fishing and shrimp trawling caused mortality of turtles.
4. Lack of political will and administrative support, coordination of management and identification of problems of the areas, etc.
5. Lacunae in publicising the negative impacts of commercial exploitation on turtles.
6. Absence of strict execution of Coastal zone regulation Act, 1992.
7. Direct and indirect impacts of tourism, beach mining and industrial pollution.
8. Lack of patrolling, combating, poaching of turtles, nests and their eggs.
9. Removal of mangroves and replacing them by aquaculture ponds.
10. There is a need to establish the main breeding grounds as Protected Areas, or as wild life sanctuaries or these turtles may be protected by launching a species project in line with the Project Tiger by involving locals and fisher folk and forest officials.

Conservation Priorities

1. Converting/encroachment of breeding grounds of turtles settlements should be banned and strictly monitored.
2. Understanding the turtle biology and protection of the fragile ranches.
3. Habitat degradation, coastal wall construction and beach nourishments should be addressed.
4. Discharge of effluents into the beaches should be checked.
5. Establishments of special conservation schemes for marine turtles and a ban on sales of turtles and their products.
6. All activities along the fragile areas should be assessed through the Environmental Impact Assessment studies.
7. Involvement of NGO's in the protection activities/ campaign.
8. Public awareness through mass media Seminars, Symposia and Workshops.
9. Ban on ecotourism or beach activities during their breeding seasons and protection by strict watch.

Pic. 32.1. A burning pipeline near the northern refinery town of Baiji

